DEVELOPMENTS IN
THIN-WALLED STRUCTURES—1

THE DEVELOPMENTS SERIES

Developments in many fields of science and technology occur at such a pace that frequently there is a long delay before information about them becomes available and usually it is inconveniently scattered among several journals.

Developments Series books overcome these disadvantages by bringing together within one cover papers dealing with the latest trends and developments in a specific field of study and publishing them within *six months* of their being written.

Many subjects are covered by the series, including food science and technology, polymer science, civil and public health engineering, pressure vessels, composite materials, concrete, building science, petroleum technology, geology, etc.

Information on other titles in the series will gladly be sent on application to the publisher.

DEVELOPMENTS IN THIN-WALLED STRUCTURES—1

Edited by

J. RHODES

*Senior Lecturer, Department of Mechanics of Materials,
University of Strathclyde, Glasgow, UK*

and

A. C. WALKER

*Professor, Department of Mechanical Engineering,
University of Surrey, Guildford, UK*

APPLIED SCIENCE PUBLISHERS
LONDON

APPLIED SCIENCE PUBLISHERS LTD
RIPPLE ROAD, BARKING, ESSEX, ENGLAND

British Library Cataloguing in Publication Data

Developments in thin-walled structures.—1.—(The
Developments series)
1. Thin-walled structure—Periodicals
I. Series
624.1'77   TA660.T5

ISBN 0-85334-123-0

WITH 7 TABLES AND 142 ILLUSTRATIONS

© APPLIED SCIENCE PUBLISHERS LTD 1982

The selection and presentation of material and the opinions expressed in this
publication are the sole responsibility of the authors concerned

Printed in Great Britain by Galliard (Printers) Ltd, Great Yarmouth

PREFACE

Many factors, including cost and weight economy in construction, development of new materials and processes and the growth of increasingly powerful methods of analysis, have contributed to the continuing increase in the use of thin-walled structures. A wide variety of structures ranging from aircraft, bridges, ships and oil rigs to storage vessels, industrial buildings and warehouses are now manufactured, at least in part, using thin-walled construction.

The research and development which underlies and stimulates the growth in the use of thin-walled structural forms has generally been reported in a variety of technical journals, usually in the form of papers or notes presenting the latest results. Space restrictions generally dictate that such papers have a length which allows the author to deal exclusively with his own work and include only brief references to the wider industrial or scientific implications of the results. These technical journals serve a very valuable purpose by ensuring that a large number of topics can be covered by the short papers. However, it is difficult in short papers to provide a general oversight into the development of any one field of activity in thin-walled structures.

In view of this, it is felt that a need exists for an outlet for papers of considerably greater length than those usual in the technical press, and which could concentrate on a review approach to the subject as well as providing the most up to date information possible. This is the underlying concept behind the present book and future volumes under the title *Developments in Thin-Walled Structures*.

It is a formidable task to attempt to present a coherent picture of the advances in all the types and applications of thin-walled structures to

v

include design, analysis and fabrication techniques for shells, plates and proprietary systems in the many areas of use. However, this very fact makes it even more worthwhile to undertake the task of editing the series, since it is only when the design engineer has a good understanding of the state of development of his particular thin-walled structure that he feels complete confidence in its use.

In the first volume we have collected seven chapters which are representative of the advances being made in the uses and analyses of thin-walled structures. We have attempted to cover a large part, but by no means all, of the broad spectrum of developments in the field. We have taken care to present developments on the design aspects and on the analytical aspects of various areas of interest.

As editors we are conscious of the slightly different styles of the various authors. We have, however, resisted the inclination to attempt to produce complete uniformity. In doing so we are assured that the reader obtains information in the form expressed by our expert contributors.

J. RHODES and
A. C. WALKER

CONTENTS

LIST OF CONTRIBUTORS

P. S. BULSON

Head of Military Vehicles and Engineering Establishment, Ministry of Defence, Barrack Road, Christchurch, Dorset BH23 2BB, UK

M. A. CRISFIELD

Senior Principal Scientific Officer, Bridges Division, Structures Department, Transport and Road Research Laboratory, Crowthorne, Berkshire RG11 6AU, UK

A. NEEDLEMAN

Associate Professor of Engineering, Brown University, Providence, Rhode Island 02912, USA

J. RHODES

Senior Lecturer, Department of Mechanics of Materials, University of Strathclyde, 75 Montrose Street, Glasgow G1 1XJ, UK

J. W. B. STARK

Head of the Department of Steel Structures, Institute TNO for Building Materials and Building Structures, PO Box 49, 2600 AA Delft, The Netherlands

A. W. TOMÀ

Research Engineer of the Department of Steel Structures, Institute
TNO for Building Materials and Building Structures, PO Box 49,
2600 AA Delft, The Netherlands

A. S. TOOTH

Reader, Department of Mechanics of Materials, University of
Strathclyde, James Weir Building, 75 Montrose Street, Glasgow
G1 1XJ, UK

V. TVERGAARD

Department of Solid Mechanics, The Technical University of Denmark,
DK-2800 Lyngby, Denmark

A. C. WALKER

Professor, Department of Mechanical Engineering, University of
Surrey, Guildford, Surrey GU2 5XH, UK

N. YAMAKI

Professor, Institute of High Speed Mechanics, Tōhoku University,
Sendai, Japan

Chapter 1

STORAGE VESSELS

ALWYN S. TOOTH

*Department of Mechanics of Materials,
University of Strathclyde, Glasgow, UK*

SUMMARY

A survey is presented of the most common bulk storage facilities used for liquid and gaseous products. The way in which the physical properties of the product to be stored have influenced the selection of the geometric shape and the vessel or tank material are discussed with particular relevance to the thin-walled storage vessel. The design features associated with these vessels are considered with reference to the relevant codes and specifications. In the main, such references are confined to British Standard specifications although mention is made of codes used in the USA. It is hoped that readers in other countries will find common ground in these references.

The aim of the chapter is to provide a background for a designer who although an expert in one area of storage may be unfamiliar with the broad picture. The list of references included should provide specialist knowledge to examine particular topics in further detail.

1. INTRODUCTION

This chapter examines the developments introduced and designs used in a wide variety of storage vessels. Although in recent years some attention has been directed to the use of underground storage, the main worldwide storage facility at present in use is the large sized thin-walled vessel. In such vessels the wall thickness may be assumed thin compared with the diameter,

1

or the length, and thus such vessels may be considered as 'thin-walled structures'. Such an assumption means that the extensive know-how available in thin-walled shell analysis may be readily used.

The predominant load in these structures is that applied by the contained product, which may be of granular, liquid or gaseous form. The designer has considerable scope in providing vessels of a particular geometry so that these loads may be carried in the most advantageous manner. In general this occurs when the stresses are uniform across the wall thickness of the vessel, i.e. when they are membrane stresses. This condition can be approximately achieved in that part of the vessel removed from discontinuities of geometry or loading. However, particular attention has to be given to those regions where discontinuities do occur, i.e. pipework connections, lifting lugs and support regions, etc., since they are invariably associated with large bending stresses. Over the years designers have considered this problem and many admirable solutions aimed at reducing these bending stresses have been provided ranging from the flexible multi-saddle support of the road tanker to the sand or bitumen cushion at the base of the spheroidal vessel.

The designer is called upon to provide storage for a very wide range of product. Some of these are non-hazardous aqueous liquids and may be stored in open ponds where clay is used to provide a watertight base. The other extreme is the land bulk storage of liquid methane at $-162\,°C$ or the storage of certain gaseous products in high pressure storage systems at 69 bar (1000 psig).

It is clear that the physical properties of the products have influenced the geometric shape of the vessels used and further, determined the type of material used to manufacture the vessel. This latter point is particularly important when vessels are required to store liquefied gases under refrigerated conditions at cryogenic temperatures.

In order to examine the way in which this has taken place and to investigate the design features of these vessels, it is proposed to classify the types of vessel in present use according to geometry rather than product.

Essentially, four different geometric shapes will be considered:

(a) flat-bottomed, vertical cylindrical tanks;
(b) horizontal cylindrical vessels with dished or formed ends;
(c) spherical or modified spherical vessels;
(d) special storage vessels.

The study will be confined to metallic vessels which are in general and widespread use.

2. FLAT-BOTTOMED VERTICAL CYLINDRICAL TANKS

By far the greatest number of vessels in use today are of the vertical cylindrical type, operating at essentially atmospheric pressure. This class of vessel can be subdivided into three broad areas according to content:

(i)   single-walled tanks with a fixed roof—generally used for the storage of non-volatile liquids;

(ii) single-walled tanks with a roof that floats with the level of the stored liquid—used where volatile liquids are to be stored;

(iii) single- or double-walled tanks used for the storage of cryogenic and low temperature refrigerated gases in liquid form.

The design of these cylindrical tanks have many features in common. For example, the determination of the wall thickness from the internal hydrostatic pressure; the derivation of the compressive axial stress (due to dead weight and wind) so that local buckling on the leeward side of the wind be avoided; the use of stiffening rings to avoid buckling in the upper sections, etc. These general aspects will, therefore, be dealt with at this point. The more specific points will be dealt with later.

2.1. General Design Features

The sizes of these tanks range from 3 m right up to 114 m diameter and for certain diameters up to 25 m in height (see BS 2654[1]). They consist of the base, shell and roof. In this section comments will be confined to the base and the single shell since these have common features for all flat-bottomed tanks.

2.1.1. The Base of the Tank (see Fig. 1(a))

The base is a flat-bottom supported on a foundation, which for the large oil storage tanks may be of sand, gravel or crushed rock, and for the smaller tanks, a concrete base. In each case the strength of the foundation must be adequate to carry the dead weight loading. Helpful advice regarding foundation requirements is given in appendix A of BS 2654[1] and CP 2004.[2] The centre part of the tank bottom consists of rectangular plates of a minimum thickness of 6 mm. These plates are lapped and welded on the top side only with a full fillet weld and with a minimum lap of five times the thickness of the plate. Where three plates meet the upper plate is hammered down prior to welding. An annular ring of approximately 500 mm width and built up of butt-welded plates is located in the region of the connection of the bottom to the vertical shell. Such a ring is obligatory for vessels greater than 12·5 m diameter. The thickness of the ring for these larger diameter tanks increases with the increase in the bottom course thickness. Details of the plate arrangement are given in BS 2654[1] and API 650.[3]

| In the case of smaller tanks and stainless steel tanks made for the food, beverage and pharmaceutical industries the base is invariably butt-welded throughout.

2.1.2. The Vertical Cylindrical Shell Region (Fig. 1(a))

This region is usually a butt-welded construction although lap joints are occasionally used—see reference 4. The plates are made up of a number of courses, or tiers, usually of the same height. Each course should be made up of a number of equal length plates. The vertical joints between shell plates should not be in alignment within any three consecutive courses and the distance between the vertical welds in adjacent courses should be one third of the shell plate length. This procedure minimises undue weld distortion and also the likelihood of a fracture following the path of a vertical weld.

To obtain the plate thickness it is assumed that the shell region removed from the base support is stressed in a membrane manner, and the discontinuity stresses which occur at a change in shell-wall thickness may be neglected. These membrane stresses arise from two sources:

(1) The hydrostatic pressure of the liquid contents which varies linearly from zero at the top of the liquid surface to a maximum at the base and is equal to specific weight, γ, multiplied by the height, h. These loads produce a circumferential stress, σ_θ.

(2) The self weight of the tank above the point under consideration, plus loads from the roof, i.e. the weight of the roof and the supporting structure, plus a minimum superimposed load of $1\cdot2\,\mathrm{kN\,m^{-2}}$ of projected area and finally from wind loads. These various loads produce a total axial stress, σ_z, in the shell.

2.1.2.1. Circumferential stress.

In general the axial stresses are small and at the design stage the plate thickness is calculated on the basis of limiting the circumferential membrane stress in the shell to the lesser of $260\,\mathrm{N\,mm^{-2}}$ and $\frac{2}{3}$ of the material specified minimum yield strength ($\mathrm{N\,mm^{-2}}$) at room temperature for all tank courses. Where the operating temperature is over $150\,^\circ\mathrm{C}$ consideration is given to the effect of that temperature on the yield strength. Values of $\frac{2}{3}$ yield strength correspond to the design stress of BS 5500[5] and are available for a range of materials at various temperatures.[5]

It is usually assumed that the worst loading case is when the tank is filled with water (specific gravity = 1·0). Since the specific gravity of crude oil is, on average, 0·86 at 16 °C, and never greater than 1·0, it means that a tank so designed may be used for any type of 'crude' with safety. When the tank is to be used for other materials of specific gravity greater than 1·0 then the correct value should be used. The relationship to determine the tank thickness is presented in BS 2654[1] and API 650.[3] This is derived from the thin shell relationship

$$t = \frac{pD}{2\sigma_\theta} = \frac{\gamma hD}{2f} \tag{1}$$

where f is the allowable or design stress.
Since γ for water = 9806 N m^{-3}

$$t = \frac{9806whD}{2f} \tag{2}$$

where w = maximum specific gravity of the contained liquid and the other dimensions are in Newtons and metres.

Equation (2) can be suitably modified to arrive at the value given in BS 2654[1]

$$t = \frac{D}{20f}[98w(H - 0\cdot3) + p] + c \tag{3}$$

where t = calculated minimum thickness (mm), H = height from bottom of course under consideration to the top of the shell (m), D = tank diameter (m), w = maximum specific gravity of the contained liquid under storage conditions, but shall not be less than 1·0, f = allowable design stress (N mm^{-2}), p = design pressure − neglected for non-pressurised tanks (mbar) (1 mbar = 100 N m^{-2}), and c = corrosion allowance (mm).

It is noted that the plate thickness is determined using the pressure 0·3 m above the centre line of the lowest horizontal joint of the plate under consideration. This is an empirical procedure which gives cognition of the fact that the lower thicker course provides some stiffening to the thinner course located immediately above. BS 2654[1] indicates that in no case shall the nominal thickness of shell plates be less than the following:

For tanks	$D < 15$ m	5 mm thickness	
For tanks	$15 \leq D < 36$ m	6 mm thickness	
For tanks	$36 \leq D \leq 60$ m	8 mm thickness	(4)
For tanks	$D > 60$ m	10 mm thickness	

These minimum thicknesses are needed for construction purposes and therefore may include any corrosion allowance. The procedure is thus to use eqn (3) for the proposed layout of courses and determine the required thickness to satisfy the design stress. If this is less that those given in the relationships of eqn (4) then the minimum values of eqn (4) should be used. Since the hydrostatic pressure varies with height the thickness of the tank shell-plate decreases in a stepwise manner upwards.

It should be pointed out that for many small tanks used for general purposes the requirements of BS 2654 regarding the minimum thickness as given in eqn (4) are unnecessarily restrictive. These vessels are shop built and since their overall size is usually less than the current limitations imposed by the Department of Transport (4·3 m wide × 27·4 m long × bridge height on route, note that motorway bridges in the UK are 5·03 m high) they may be transported to the site in a completed condition. Readers familiar with older codes will note that eqn (3) does not contain a joint efficiency factor, introduced because of the likelihood of weld defects in tanks which were only visually inspected. Since 1965 a joint efficiency of 1·0 is used. This is associated with welding procedures and radiographic inspections as laid down in BS 2654.[1]

2.1.2.2. The axial stress. Once the wall thickness has been determined it is necessary to estimate the axial stress in the shell. In the event that this is above the allowable value then the wall thickness has to be increased. The axial stress arises from the total dead weight, W, plus that due to an overall bending moment, M, caused, most often, by the wind, that is

$$\sigma_z = \frac{W}{\pi D t} \pm \frac{4M}{\pi t D^2} \tag{5}$$

It should be noted that the moment, M, produces a stress, σ_z, which is approximately uniform across the wall thickness. On the windward side this axial stress is tensile and on the leeward it is compressive. It is noted that eqn (5) is based on the 'Engineers Bending Theory' where the circular shape is assumed to only undergo small displacements. This is considered a reasonable assumption in that the maximum axial stress often occurs at the base, where the shape is restrained. The value of M is determined from the wind loading on the tank. To do this it is necessary to determine:

(a)   The geographical location of the vessel and from this the basic wind speed, V, which is the 3-s gust speed estimated to be exceeded on average once in 50 years.

(b)   Three wind speed factors, S_1, S_2 and S_3 which enable the topography, ground roughness and a freak wind probability factor to be considered.

Values of the wind speed V for UK locations and values for the above factors are given in British Standards (see reference 6). For areas outwith the UK the wind speed information can be obtained from local meteorological stations.

The design wind speed V_s is given by

$$V_s = VS_1S_2S_3 \tag{6}$$

This is converted to a dynamic pressure by using

$$q = \tfrac{1}{2}\rho V_s^2 \tag{7}$$

where ρ is the density of air. The figure used is the density of air at 15 °C and under atmospheric pressure, viz. $\rho = 1\cdot227\,\mathrm{kg\,m^{-3}}$. Thus,

$$q = 0\cdot613V_s^2 \qquad (\mathrm{N\,m^{-2}}) \tag{8}$$

The pressure varies round the tank in such a way that on the windward side only the circumference of the tank covered by $\pm40°$ is subject to a radial inward pressure. The rest of the tank is subject to suction, i.e. an outward pressure. Details of this variation are given in reference 6. In view of this variation the total horizontal wind load on the shell is given by

$$F = C_f q A_e \tag{9}$$

where C_f is the force coefficient for the tank and takes into consideration the pressure variation. It varies from $0\cdot5$ to $1\cdot2$ depending upon the height/ diameter ratio, the velocity of the wind and the smoothness of the tank, i.e. pipe projections, etc. (see reference 6), and A_e is the effective frontal area, i.e. the area normal to the wind.

It is generally assumed that the dynamic wind pressure is constant with the height of the vessel so that the resultant wind force, F, acts at mid height or alternatively it may be considered as a uniformly distributed force up the shell. Using this assumption the overall moment, M, can be determined.

Since the wall thickness varies it is necessary to check the value of σ_z as obtained from eqn (5) at different heights. On the tensile, windward side the allowable stress for the non-pressurised tank is that given above, viz. the lesser of $260\,\mathrm{N\,mm^{-2}}$ and $\tfrac{2}{3}$ yield stress. However, on the compressive leeward side it is important to consider the likelihood of wrinkling or local buckling occurring. The classical value of the critical buckling stress for the end-loaded ideally circular shell of radius R is given by

that is,
$$\sigma_c = Et/\sqrt{3(1 - v^2)}R \tag{10}$$
$$\sigma_c = 0\cdot605Et/R \quad \text{if} \quad v = 0\cdot3$$

For the actual shell manufactured to engineering tolerances eqn (10) is very optimistic since it does not allow for shell imperfection, nor does it incorporate any factor of safety.

Allowable axial compressive stress. A number of equations have been proposed to determine the allowable compressive stress in actual tanks. That by Brownell and Young[4] for steel and in imperial units is

$$\sigma_{c\ (allowable)} = 1\cdot5 \times 10^6 \left(\frac{t}{R}\right) \le \tfrac{1}{3}\sigma_{yield} \tag{11}$$

Assuming $E = 30 \times 10^6\ \text{lbf in}^{-2}$, eqn (11) may be written

$$\sigma_{c\ (allowable)} = 0\cdot05E\frac{t}{R} \le \tfrac{1}{3}\sigma_{yield} \tag{12}$$

That contained in BS 1500[7] is a simple relationship

$$\sigma_{c\ (allowable)} = 0\cdot0625\frac{Et}{R} \tag{13}$$

Equation (13) was modified in BS 1515[8] to consider the 'unevenness' of the tank or vessel. The more recent code for pressurised containers, BS 5500,[5] gives a comprehensive treatment of vessels subject to external pressure. This can be used for cylinders where the longitudinal stress is the governing condition; i.e. axial loading or bending producing axial stress, and when the tank, or vessel, is not subject to external pressure. In this code the allowable stress is limited to $\sigma_c = \Delta sf$, where f is the design stress; $s = 1\cdot4$ for carbon, carbon manganese and ferritic alloy steels and $1\cdot1$ for austenitic steels and aluminium alloys; Δ is obtained from a curve of $K \sim \Delta$, where

$$K = 0\cdot605Et/sfR \tag{14}$$

The value of Δ can also be obtained from the following empirical relations:

$$\Delta = 0\cdot5[1 - (1 - 0\cdot125K)^2] \quad \text{for } K \le 8$$
$$\Delta = 0\cdot45 + 0\cdot00625K \quad \text{for } 24 \ge K \ge 8$$
$$\Delta = 0\cdot6 \quad \text{for } K \ge 24$$

It is found that the most conservative allowable buckling stress (i.e. the lowest) is given by:

1. $0.05Et/R$ in the large R/t range ($R/t \geq 300$);
2. BS 5500 for lower R/t values, when the design stress f is on the low side ($f \leq 120 \text{ N mm}^{-2}$).

However, the differences are not large and it is concluded that since BS 5500 incorporates, in the $\Delta \sim K$ variation, a very large number of experimental results and considerable operating experience the procedure of BS 5500 is to be preferred and should be used to obtain the allowable compressive stress.

The values of σ_z obtained from eqn (5) should be compared with this allowable stress and where required plate thicknesses should be increased.

2.1.3. Shell to Base Connection
The vertical shell is welded to the base plates by means of a double fillet weld, that is, a weld on either side of the shell. During filling and emptying this joint prevents the shell from moving radially and, to a more limited extent, from rotating relative to the base. The restriction offered by the tank bottom results in bending stress in the shell, the weld region and the base plate. During the continual filling and emptying this can result in a high cyclic stress likely to cause failure. An example of such a failure is given by Cotton and Denham.[9] This shows a crack initiating at the toe of the weld in the base plate and propagating through the plate.

The effect of the high stress is compounded since this region is often contaminated by unwanted spillage from the tank. Corrosion cracking can occur in a short period if the weld hardness exceeds a threshold value. In such cases the region should be protected with tar epoxy paints, against deleterious spillage products. It is therefore important that a good quality connection be made in this region. To achieve this the following are considered of value:

1. that the shell plate receive a weld preparation for a double fillet weld;
2. that backgrinding (or gouging) take place on the reverse side of the first fillet weld, prior to welding of that side, so that an exposed surface be left for proper interfusion.

To determine analytically the magnitude of the bending stress in the shell/base region, it is necessary to examine the behaviour of the shell and base. A simple approach is to assume that the base is rigid throughout and only the shell deforms. The resulting bending moment obtained by this approach will be an over-estimate of the actual value since in practice the base will rotate slightly and relieve the end fixing moment. The relationship presented by Flügge[10] based on the shell equations is very similar to that of Den Hartog[11] using the beam on elastic foundation analogy. That of Flügge is given by

$$M_z = -\frac{\gamma R t}{\sqrt{12(1 - v^2)}} \left[\frac{R}{\chi} - h\right]$$

where

$$\chi = [3(1 - v^2)R^2/t^2]^{1/4} \qquad (15)$$

and h is the height of the liquid in the tank.

From this the axial stress in the shell wall at the base can be determined, viz.

$$\sigma_z = \pm \frac{6M_z}{t^2} \qquad (16)$$

It should be noted that unlike the overall turning moment M given in eqn (5) which produces an essentially membrane axial stress, M_z produces a stress which varies across the wall thickness. The sign of the moment obtained from eqn (15) is such that when positive it produces a compressive stress on the outer plate surface and a tensile stress on the inner surface. Using the value of this stress it is possible to make an assessment of the fatigue behaviour of the joint and thus be alerted to the potential danger in this region of the tank.

2.1.4. The Stability of the Tank—Wind and Vacuum

During high wind loading both the floating-roof and fixed-roof tanks are likely to buckle. In the case of the floating-roof tank this could cause a major collapse of the tank since, without the provision of wind girders, it lacks the stabilising effect of the roof. Both types are likely to be subject to local buckling of some of the thinner-walled panels, which although not as serious as that referred to earlier is costly to repair and furthermore indicative of a vessel which is underdesigned. Under high wind conditions, critical conditions occur, (i) when the tank is empty and therefore lacking the beneficial effect of the internal hydrostatic pressure, (ii) in a fixed-roof tank during emptying or during evaporation, both of which cause a vacuum, and (iii) during construction prior to the installation of the fixed roof, or other ring girders. In this respect the construction procedure whereby the tank is fabricated roof first at ground level and jacked up tier by tier, is interesting and should avoid this problem.

Extensive investigations into the stability of these tanks have been carried out over many years. Although some progress has been made to provide an analytical approach to the problem,[12-18] the contribution by De Wit[19] in providing a design procedure must be recognised. This approach serves as the basis of that presented in BS 2654.[1]

2.1.4.1. Primary wind girders. In the case of a closed fixed-roof tank the wind load is only external, whereas in a floating-roof tank the wind loading also acts on the inner surface and on that surface may be assumed to behave like a vacuum load. The fixed roof assists in maintaining the overall rigidity of the shell so that it is capable of transmitting the wind forces to the foundation by means of the axial stresses referred to earlier. Tanks with floating roofs lack this rigidity and it is therefore necessary to provide a wind girder round the whole circumference. This is referred to as a primary wind girder and is located at or near the top of the top course and preferably on the outside of the tank shell.

The required minimum section modulus of the girder is determined from the equation (in cm^3)

$$Z = 0.058D^2H \tag{17}$$

where the tank diameter D and height H are in metres.

As pointed out by De Wit,[19] 'This formula has been used for many years for the calculation of the strength of wind girders. It is an approximate relationship worked out at a time when the vessels under construction were less than 30 m in diameter'. Reference to API 650[3] indicates that the formula is based on a wind speed of 100 m.p.h., values for other wind speeds being obtained by multiplying the equation by $(V/100)^2$. The formula may be derived by using this wind speed (100 m.p.h.) and obtaining the dynamic wind pressure from eqn (8). The total horizontal wind load, in terms of D and H, can be obtained from eqn (9) using a C_f value of 0·6. Assuming that the ring is loaded with a uniform pressure across the diameter and supported by tangential shear, and that the pressure load on the top 25 % of the tank has to be taken by the wind girder and the allowable design stress is 103·42 N mm^{-2} (15 000 psi), which is increased by 25 % because it is a wind load, the value for the section modulus can be shown to be given by eqn (17).

Further discussion about eqn (17) may be found, by Adams[20] who considers that those vessels where the shell weight equals, or exceeds, the maximum wind induced uplift on the shell do not need wind girders to prevent the shells from blowing in. He considers that only nominal sized rings for stabilisation are necessary.

It has been found by many that for vessels with diameters greater than 60 m eqn (17) predicts unnecessarily strong girders and therefore the values of D are held constant at 60 m for these tanks when using eqn (17). Design details for a range of these girders are given in BS 2654.[1]

2.1.4.2. Secondary wind girders. Both open top and fixed roof tanks may require secondary wind girders to maintain roundness over the full height and prevent local buckling of the shell. The number and location of these stiffening girders depends upon the wind loading and the vacuum conditions in the tank. Vacuum is assumed to occur in the open top (floating-roof) vessel during wind loading and although the actual distribution is not uniform it is assumed to be 5 mbar irrespective of wind speed. Many fixed roof tanks are fitted with pressure/vacuum valves which are set to open when the vacuum in the tank reaches a certain value, say 2·5 mbar in a non-pressure roof. The reason for using these valves is to contain the vapour losses which take place during evaporation. However, by the time the valve is fully open and developing the full design throughput the vacuum will have increased by about 2·5 mbar to a total of, say 5 mbar. For other fixed-roof tanks a total of 8·5 mbar vacuum is recommended (see BS 2654).[1]

The dynamic wind pressure on the tanks is obtained from eqn (8) (in Nm^{-2})

$$q = 0·613V_s^2$$

To this must be added the vacuum, V_a (mbar). The total loading of wind plus vacuum is therefore (in Nm^{-2})

$$q = 0·613V_s^2 + 100V_a \tag{18}$$

The design procedure set out in references 19 and 1 makes the assumption that the shell is subject to a uniform external pressure and is modified to a constant wall thickness. The actual shell with varying plate thickness is converted into a hypothetical shell with a diameter and wall thickness equal to that of the top course with the height reduced in such a way that the stability of the actual shell is equal to that of the hypothetical shell. To justify the approach reference can be made to the work of Saunders and

Windenberg,[21] who present an approximate relationship for the uniform external pressure of q at which elastic buckling occurs in a short tube, of length l, with ends held circular, or a long tube held circular at intervals l. Their relationships have been simplified by Roark[22] and may be written

$$q' = \frac{0 \cdot 807E}{l} \left[\frac{1}{(1 - v^2)} \right]^{3/4} \frac{t^{5/2}}{R^{3/2}} \tag{19}$$

Keeping R constant in eqn (19) an equivalent buckling pressure q' is achieved when $l \propto t^{5/2}$. Thus,

$$H_e = h_t \left(\frac{t_{min}}{t} \right)^{5/2} \tag{20}$$

where t_{min} = thickness of top course (mm), t = thickness of each course in turn (mm), h_t = height of each course in turn below the primary girder (m) and H_e = equivalent height of each course at thickness t_{min} (m).

The total height H_E of the hypothetical shell is obtained by adding together the heights of all the hypothetical shell courses H_e, i.e.

$$H_E = \Sigma H_e \tag{21}$$

By equating the actual pressure q (eqn (18)) with the pressure to cause buckling q' (eqn (19)) it is possible to determine a value for the maximum permitted spacing, l, of the secondary rings on this uniform thickness shell—designated H_p (m) in BS 2654[1]

$$l = H_p = \frac{0 \cdot 807E}{0 \cdot 613V_s^2 + 100V_a} \left[\frac{1}{1 - v^2} \right]^{3/4} \left(\frac{t_{min}}{1000} \right)^{5/2} \left(\frac{1}{R} \right)^{3/2}$$

It is found that using E and v for steel the above reduces approximately to the relationship given in BS 2654[1]

$$H_p = \frac{95\,000}{3 \cdot 563V_s^2 + 580V_a} \left(\frac{t_{min}^5}{D^3} \right)^{1/2} \tag{22}$$

If $H_p > H_E$ then the shell is sufficiently stable and does not require any secondary rings. If $H_p < H_E$ then one or more rings are necessary. In many cases these rings will be located on the top course, or on a course of similar thickness, so that they are located in the same place on the actual shell as on the hypothetical shell. If a secondary ring has to be located lower down, its correct location on the actual course must be determined by converting the equivalent heights back to the actual heights. The whole procedure is illustrated by examples in BS 2654[1] and reference 19. The actual size of the stiffeners given in reference 1 is not related to design loads but only to tank diameter. They only serve to prevent local shell buckling and need not be large. They range from an angle section of $100 \times 60 \times 8$ mm for vessels less than 20 m diameter to $200 \times 100 \times 12$ mm for vessels of 48 m diameter and above. A more precise method for determining the location and size of these stiffeners is given in BS 5500.[5] However, it would appear from experience that these vessels are adequately designed to resist vacuum and wind loading (see Adams[20]).

2.2. Single-Walled Shell With a Fixed Roof

These tanks are probably the most widely used and provide the cheapest non-pressure storage at normal temperatures, i.e. outwith the cryogenic and creep ranges. The smaller diameter tanks are widely used in the food, beverage and pharmaceutical industries, where mixers are fitted to the tanks to provide continuous stirring of the product. The effect of these is to induce vibrations which can produce large tank displacements if their frequency of vibration is in sympathy with one of the natural frequencies of the tank. Although throughout the tank the stress level of these effects may be small, certain welds may be highly stressed, e.g. the toe of a fillet weld; such often provide the initiation point for a propagating crack. A similar effect has been noted originating from a pumping unit. In this case the pump vibrations are transferred through the supply pipe which has been fixed with brackets to the tank wall. Where the vibrations cannot be isolated it is a worthwhile exercise to design the tank—using, say, additional ring stiffening—so that the natural frequencies of the tank are removed from the rated speed of the mixer unit. References 23, 24 and 25 have been found useful in this respect.

The above is not usually a problem in the case of the larger tanks, many of which are used for storing oil and petroleum products. Here considerable thought has been given to the control of the vapour which is located in the space between the contents and the roof. This is a most valuable commodity and losses should be kept to a minimum. Much of the loss occurs due to the variation of night and day temperatures. The hot day temperature causes evaporation and expulsion of vapour through the vents. At night the reverse occurs and air is drawn into the vessel. These breathing losses can be reduced by installing pressure/vacuum valves in the roof. They are set to prescribed limits so that under normal ambient temperature variations, say a range of 8–18 °C, breathing does not occur. Losses can also occur during filling when the inflowing oil displaces an equal quantity of vapour. As will be noted later many of these problems are overcome by using a floating roof. The general design features detailed in

Section 2.1 are relevant to the base and shell of the single-walled fixed-roof tank. Comment here is therefore confined to the roof structure and plating. There are basically three types of roof in common use:

1. The self-supporting structure.
2. The column support structure.
3. The frameless roof.

2.2.1. The Self-Supporting Roof Structure

This type of roof consists of a structure which spans the tank diameter with a series of radial arms and is overplated (or in some cases underplated) with roof sheeting. The structure may be in the form of a trussed, coned or domed roof as shown in Fig. 1((a), (b) and (c)), designed to support the superimposed load, which has a minimum of $1 \cdot 2 \, \text{kN m}^{-2}$ of projected area. The slope of a cone roof is usually 1 in 5 and the domed roof has a radius of curvature in the range $0 \cdot 8 - 1 \cdot 5D$ (where D is the tank diameter). The cone roof may be designed as a series of 3-pinned arches and the domed roof as a series of 2-pinned arches. The load on each arm is that taken by the roof plate sector immediately above and midway between the adjoining arms. This 'arm' loading can be idealised by assuming a distributed load which varies linearly from maximum at the outside to zero at the centre line of the tank. In assuming that the arms behave as arches it is appreciated that the top of the vertical shell must be capable of applying horizontal inward acting reactive forces. To assist in carrying these forces a top curb angle is generally used. Sizes of these are given in BS 2654[1] and API 650,[3] but their strength should be checked assuming they carry the horizontal forces calculated from the arch calculations. These forces produce a tensile circumferential stress in the curb angle.

When internal pressure is acting in the tank the region of the shell roof junction is subject to compression. By considering the forces in this region, it is possible to find the required area to avoid high compressive stresses occurring. The vertical reactive force at the shell roof junction $= pR/2$ per unit length of periphery.

Assuming only membrane forces at this point, the horizontal force at the shell roof junction equals $pR/2 \tan \theta$ per unit length where θ is the slope of the roof meridian (see Fig. 1(b)).

Considering this region as a ring subject to $pR/2 \tan \theta$ per unit length of periphery results, using circumferential equilibrium, in a circumferential compressive stress of

$$\sigma_\theta = pR^2/2A \tan \theta \tag{23}$$

where A is the area of the region.

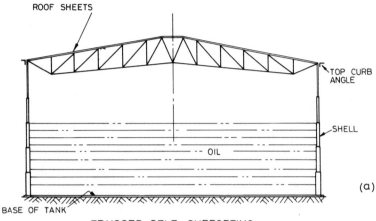

TRUSSED-SELF SUPPORTING

(a)

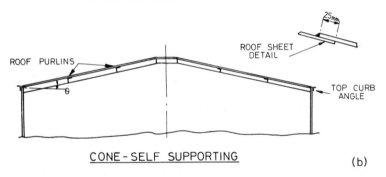

CONE - SELF SUPPORTING

(b)

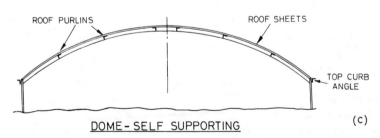

DOME - SELF SUPPORTING

(c)

FIG. 1. Self-supporting fixed roofs.

If the circumferential stress be equated to the allowable stress in compression σ_c then the area provided shall not be less than that given by

$$A = 50pR^2/\sigma_c \tan \theta \qquad (24)$$

where A = area required (mm^2), p = internal pressure (less the weight of the roof sheets) (mbar), R = radius of the tank (m), σ_c = allowable compressive stress (which unless otherwise specified shall be taken as 120 N mm^{-2}).

The required area A is the cross-sectional area of the curb angle plus certain lengths of the shell and roof considered to be close enough to the junction to be effective. Details of these lengths are shown in BS 2654.[1]

The radial 'arms' are connected together by a series of roof purlins, which run in a circumferential manner, to provide an integrated structure. The structure shall be designed in accordance with BS 449[26] following the suggested spacing for roof purlins given in BS 2654.[1]

The roof plates, which have a minimum thickness of 5 mm are lap-welded on top side only with a minimum lap of 25 mm. They are lapped so that the lower edge of the plate nearest the crown is beneath the upper edge of the plate one step down (see insert in Fig. 1(b)) thus avoiding the risk of condensed moisture becoming trapped in the lap joint. (In the case of smaller tanks it is usual to butt weld throughout.) The roof sheets are not usually attached to the roof structure, but only to the top curb angle where a continuously applied fillet seal weld is used. The tank is thus free to breathe with the changes in temperature and barometric pressure or in fact to blow out in the event of an explosion in the tank with no damage to the tank shell and the containment of the product.

2.2.2. Column-Supported Roof Structure

In this case a shallow coned roof (slope = 1 in 16) is used. It is supported at regular intervals and over the total area by a series of vertical columns. Starting with a central post these are located on a sequence of equispaced concentric circles. The spacing of the columns, both radially and circumferentially, is such that straight beams can be placed circumferentially on the columns and lighter radial members span the beams. Figure 2 shows this type of roof under construction. The method is more economical for very large vessels, say over 60 m diameter, than that of the self-supporting roof structure previously discussed. It is not recommended, however, that this type of roof be used on foundations where differential settlement is anticipated. Nevertheless the worst effects of such settlement can be avoided if the base of the columns are not fixed to the foundation, but merely located in a 'box' arrangement of angle irons which allows for the

Fig. 2. Column supported roof under construction. (Courtesy, Motherwell Bridge Eng., Ltd)

relative vertical movement of the column and base. If settlement does occur the base of the vessel is thus allowed to displace without undue transfer of load to the roof and shell walls. The total roof structure is highly redundant and the fact that some of the columns do not carry any load does not unduly affect the overall performance of the roof structure.

As in the previous case the roof plates are single lap-welded together and not welded to the roof structure but to the top curb angle with a seal weld.

2.2.3. Frameless Roof
For fixed roofs up to and including 12 m diameter it is possible to construct the roof without a supporting structure, that is the plating itself has adequate structural strength. These are referred to as 'membrane roofs'. The plates are of butt-welded or double lap-welded construction. It is possible to add to their strength by folding the edges of alternate plates prior to welding, thus creating a stiffened plate structure. BS 2654[1] suggests that the roofs be designed to resist buckling due to external loading and be checked for internal pressure.

2.2.3.1. Internal pressure. For internal pressure the simple membrane stress is assumed to act, which for spherically domed roof of radius of curvature R_1 gives a roof plate thickness, t_r, of

$$t_r = pR_1/2f \qquad (25)$$

Equation (25) can be written

$$t_r = pR_1/20f\eta \qquad (26)$$

where $R_1 = $ the radius of curvature of the dome (m), $p = $ the internal pressure (mbar), $f = $ the allowable design stress (N mm^{-2}) and $\eta = $ the joint efficiency, which varies from $1\cdot0$ for butt joints to $0\cdot35$ for lapped joints (see reference 1).

For a conical roof again under internal pressure the maximum membrane stress occurs in the circumferential direction at the shell roof junction, i.e.

$$\sigma_\theta = pR/t_r \sin \theta \qquad (27)$$

where θ is the slope of the conical roof (see Fig. 1(b)). Equation (27) can be written

$$t_r = pR/10f\eta \sin \theta \qquad (28)$$

where p, f and η are defined as above.

2.2.3.2. External pressure. It is also necessary to design these roofs to withstand the effect of external pressure—caused by roof loading and by vacuum. In the case of the domed roof the classical value of buckling pressure for the perfect sphere is given by

$$q' = 2Et_r^2/R_1^2\sqrt{3(1 - v^2)} \qquad (29)$$

Using a value of $v = 0\cdot3$, eqn (29) may be written

$$q' = 1\cdot210Et_r^2/R_1^2 \qquad (30)$$

As pointed out in the previous Section 2.1.2.2. these classical results do not allow for vessel imperfections, nor factors of safety. In this case a figure of approximately $1/20$ for eqn (30) has been used in BS 2654,[1] viz.

$$allowable\ external\ pressure = 0\cdot0625Et_r^2/R_1^2 \qquad (31)$$

The stress in the spherical membrane roof is given by $P_eR_1/2t_r$, where P_e is the external roof pressure loading. By equating P_e to eqn (31)

$$allowable\ stress = 0\cdot0625E\frac{t_r^2}{R_1^2}\frac{R_1}{2t_r} = 0\cdot031\,25E\frac{t_r}{R_1} \qquad (32)$$

Equation (32) indicates that a lower factor is proposed here than that of eqn (12) for the vertical shell. This is not unreasonable since the membrane sphere is equally stressed throughout. Equation (31) may be rearranged to give the allowable plate thickness

$$t_r = 40R_1 \sqrt{\frac{10P_e}{E}} \tag{33}$$

where R_1 = radius of curvature of the domed roof (m), P_e = external loading (kN m^{-2}), E = Young's modulus (N mm^{-2}) and t_r = roof plate thickness (mm).

For a conical roof subject to external pressure the procedure given by Brownell and Young[4] is used. The allowable stress of eqn (32) is modified for a conical roof by putting $R_1 = R/\sin\theta$ (i.e. the radius of curvature of the cone at the shell–cone junction, (see Fig. 1(b)). Thus eqn (32) may be written

$$allowable\ stress = 0.031\,25Et_r \frac{\sin\theta}{R} \tag{34}$$

Equating this to the actual maximum membrane stress in the roof at the shell–cone junction, i.e. eqn (27), and substituting for the pressure p the value P_e which is the external roof loading, the following expression for the allowable plate thickness is obtained:

$$t_r = \frac{40R}{\sin\theta} \sqrt{\frac{20P_e}{E}} \tag{35}$$

where R = the vessel radius (m), P_e = the external loading (kN m^{-2}), E = Young's modulus (N mm^{-2}), θ = the slope of the conical roof and t_r = roof plate thickness (mm).

2.3. Single-Walled Shell With a Floating Roof

This is a standard tank with a butt-welded shell which is open at the top where the roof floats directly on the liquid of the tank and blankets the surface. This technique reduces the following: evaporation loss and thus product loss; air pollution, since the vapour is contained; fire and explosion risk, since the vapour space is at a minimum; and vapour losses during filling. The seal between the shell and roof almost eliminates the escape of vapour around the rim and the roof. It is contended that the extra cost of a floating roof tank over 20 m diameter can be recovered in two to three years under average working conditions. Such tanks are used for the storage of crude petroleum, finished gasolines and some of the less volatile natural gasolines.

The general design features detailed above in Section 2.1, viz. wind, stability, etc., are relevant to the base and the open top shells used in this type of storage. Thus comment here is confined to the floating-roof structure.

Essentially two forms of structure are available for the roof: (a) single-deck pontoon roofs and (b) double-deck roofs; and two types of roof seal: metallic and fabric.

2.3.1. Single-Deck Pontoon Roofs

The roof floats as a disc on the oil and rises and falls with the oil level. The pontoon-type roof is one having a continuous annular pontoon, divided by bulkheads into liquid tight pontoon compartments.

The central area is covered by a single-deck diaphragm made up of 5 mm thick lap-welded plates welded to the inner side of the pontoon. The top of the pontoon ring is sloped downwards towards the centre of the roof to facilitate drainage. The bottom deck of the pontoon is pitched upward toward the centre to deflect vapours formed under the pontoon towards the centre of the roof. This will balloon upward to provide additional space for vapours formed due to increase in temperature. The roof is shown diagrammatically in Fig. 3(a). As an alternative, on larger vessels, say over 20 m diameter, a centre pontoon is provided. This gives extra buoyancy and also reinforces the centre section of the roof.

In both the single- and double-deck roofs the disposal of rainwater is an obvious requirement. This is discharged from the roof by means of an articulated pipe drain in the tank which follows the level of the roof. The importance of adequate maintenance of this drainage system cannot be over-emphasised since some roof failures have been traced to malfunctioning exit valves. Access to the roof is by ladder which automatically adjusts for roof height so that even the stair treads remain in the horizontal position. To prevent the roof bottoming and fouling access pipes, etc., on the bottom, and also to provide under roof access, vertical leg roof supports are provided.

The buoyancy for this case is supplied by the pontoon, which covers approximately 20–25 % of the total roof area. The code, BS 2654,[1] stipulates that the minimum pontoon volume shall be sufficient to keep the roof floating on a liquid with a specific gravity not exceeding 0·7 if the single-deck and two pontoon compartments are punctured and the primary roof drain is considered as inoperative. The code also requires the

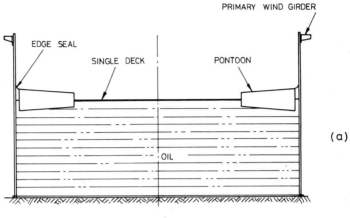

SINGLE DECK PONTOON ROOF

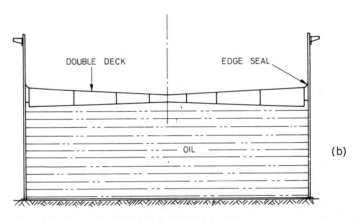

DOUBLE DECK ROOF

FIG. 3.   Floating roofs.

buoyancy to be able to handle a load of 250 mm rainfall over the entire roof area.

Some difficulty has been experienced from the cracking of the single-deck around the leg supports. This has been attributed to bending stresses caused by frequent grounding of the roof and vibration set up in the roof by certain wind conditions. Cotton and Denham[9] make mention of severe deck cracking, along the main lap-weld seams as well as in the leg support region of several floating-roof tanks located in the South of France. The cause of these failures has been attributed to the famous Mistral wind. For this reason they suggest the single deck roof be limited to 58 m diameter and various forms of circumferential and radial stiffeners be used to avoid the failures. Radial box girder stiffeners have been used together with a butt-welded centre-deck. This approach provides a completely rigid diaphragm with a reduced number of supports which are further designed to eliminate vibration. Such roofs have been used up to about 88 m diameter[9] and appear to be satisfactory.

2.3.2. Double-Deck Floating Roof

This roof was introduced to avoid some of the problems referred to above. It has a double deck over the entire liquid surface and divided into a number of compartments by circular rims and radial bulkheads to provide maximum stability. The top deck is sloped to the centre uniformly for drainage and the bottom deck pitched upward—as shown diagrammatically in Fig. 3(b). The roof has a minimum of approximately 400 mm of dead air space between the top and bottom decks which provides insulation and reduces the rate of heat transfer to the stored product. The buoyancy requirements of BS 2654[1] referred to above must also be complied with in this case.

2.3.3. Roof Sealing

To enable the free movement of the floating roof up and down the shell a flexible roof seal is installed between the roof and the shell. These are essentially of two types, metallic and fabric.

2.3.3.1. The metallic seal.

The metallic seal with pantagraph hangers are the most widely used. They consist of galvanised steel, stainless steel or aluminium sealing rings, the bottom of which remain below the liquid surface. A continuous vapour-tight and weatherproof synthetic fabric is used to close the space between the sealing ring and the rim of the floating roof. The sealing ring is supported and held firmly against the vessel by a series of pantograph hangers which apply a uniform outward radial pressure (see Fig. 4(a)). The one shown incorporates a counter weight. An alternative to this is a spring system. These pantograph hangers will accommodate a rim space variation of ± 125 mm. To do this, provision is included for expansion and contraction of the sealing ring either with vertical flexures or an 'inter-shoe' seal.

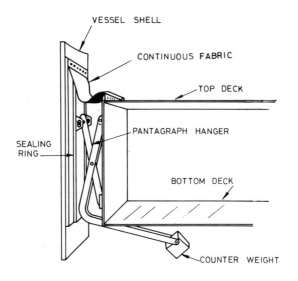

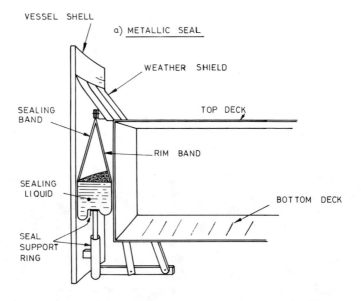

FIG. 4.   The floading roof scaling devices: sectional view. (Courtesy, Motherwell
Bridge Eng., Ltd.)

2.3.3.2. The fabric seal. A range of these is available, namely liquid-filled, as shown in Fig. 4(b), gas-inflated and a resilient foam-filled seal. With these a metallic interleaved weather-shield is usually included. It is claimed that the fabric seal provides the closest fitting seal available with a gentle non-abrasive contact with the shell, thus avoiding deterioration of any corrosion-resistant coatings on the shell. In the liquid-filled seal shown in Fig. 4(b) two continuous bands of fabric, coated with petroleum-resistant synthetic rubber, are vulcanised together at their base to form a liquid tight pocket. The tube is filled to a depth of 200–250 mm with sealing liquid and positioned within the rim space by a continuous bottom ring. The sealing bands are made of nylon cloth coated with hard-wearing synthetic rubber. In the case of the inflated seal tube a small pressure of about 4 mbar is maintained in an elliptically-shaped tube by means of a small gasholder located inside one of the roof's pontoons. As before the tube is located within the rim space. This type has the advantage that there is automatic pressure control and replenishment of gas within the tube. Urethane foam, rather than gas or liquid pressure is used in the resilient foam-filled seal to provide the sealing. This has the advantage over the liquid- and gas-filled types in that small tears or abrasions in the seal will not cause sudden failure.

On the face of it the fabric seal has all the elements of a very simple system without the problems of the more complicated pantograph mechanism. Despite this the metallic seal and pantograph is still widely used since it is considered more reliable over the long periods of time involved.

Before concluding this section it should be pointed out that certain manufacturers offer a closed fixed-roof tank within which they incorporate a floating roof, thus providing the best of both worlds. Since the most severe weather conditions do not reach the floating roof no elaborate drainage is required. There are thus considerable savings in the maintenance of the roof although initial capital costs are high. Design of these internal covers are dealt with in AMD 1750 (August 1975) of BS 2654.[1]

2.4. Single- and Double-Walled Shells Used for Cryogenic and Low Temperature Storage

Unlike the crude oil and petroleum products referred to above, which can be stored at ambient temperatures and virtually atmospheric pressure, the wide range of liquid gas products (butane, propane, methane, etc.) can only be stored in liquid form under either pressure, refrigeration or a combination of both. The reason for this is that their vapour pressure at atmospheric temperature is in excess of atmospheric pressure. It is further noted that the critical temperatures of ethane, methane (LNG), oxygen and nitrogen are below normal ambient temperatures and therefore they cannot be liquefied by pressure alone and must be refrigerated. After refrigeration they are stored at essentially atmospheric pressure. The temperature of refrigeration must be below the boiling temperature of the liquid concerned; for example, nitrogen is stored at $-196\,^{\circ}$C and methane at $-162\,^{\circ}$C. On the other hand, butane and propane, known as liquid petroleum gas (LPG), can be liquefied under pressure or refrigeration and pressure. However when such pressure vessels get very large they do represent a much greater hazard in the event of vessel failure than the corresponding refrigerated cylindrical tank form of storage. There is a tendency, therefore, in the case of large bulk storage to store all of these products in vertical cylindrical tanks, adequately insulated, in association with refrigerated units.

Single-walled tanks, with external insulation, have been widely used to store LPG. Such tanks are generally restricted to operating at $-50\,^{\circ}$C and are covered by BS 4741[27] and API 620 Appendix R.[28] The double-walled tanks, with interface insulation, operate down to $-196\,^{\circ}$C. They are covered by BS 5387[29] and API 620, Appendix Q[28] and are used for LNG storage. Over the years the size of these storage tanks has increased. A typical size now being used by British Gas[30] as part of their 'LNG Peak Shaving Programme' is 50 000 m³ with a vessel size of 46 m inside diameter and 32·6 m overall shell height. A diagrammatic sketch of such a tank is given in Fig. 5. It should be pointed out, however, that vessels twice this capacity have been successfully built and are in operation outside the UK.

Like the tanks previously discussed, these tanks must be capable of sustaining the loadings referred to earlier, in the general design features section, i.e. dead weight and wind, etc. (see Section 2.1), together with the additional factors because of the thermal loadings.

2.4.1. Insulation

It is necessary in these tanks to use insulating material extensively in the foundation, shell wall and roof. Material in the foundation must be capable of load bearing, while for the double-walled tank the interface between the inner and outer tanks makes use of a perlite infill and requires a fibrous blanket against the inner tank. The reason for this is that with repeated thermal cycling of the inner tank the material progressively compacts, building up an external pressure on the inner tank. The blanket accommodates the movement and prevents compaction. This topic has received

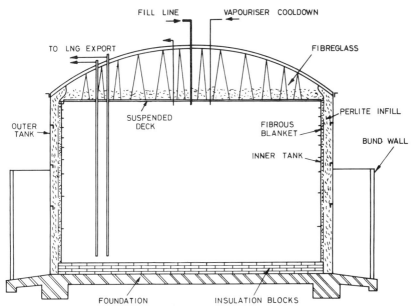

FILL LINE   VAPOURISER COOLDOWN

TO LNG EXPORT   FIBREGLASS

OUTER
TANK

SUSPENDED
DECK

FIBROUS
BLANKET

INNER TANK

PERLITE INFILL

BUND WALL

FOUNDATION   INSULATION BLOCKS

FIG. 5.   Diagrammatic cross-section of double-walled tank for LNG.

considerable attention and readers are referred to Carter and Rowley[31] and
Wardale.[32] It is interesting to note that[32] the cost of insulation for these
large vessels is roughly one third of the total cost of the whole facility and
thus there is clearly scope for the development of new materials and
concepts designed to bring about a reduction in the overall cost.

2.4.2. Tank Roofs

Because the liquids stored are at, or close to, their boiling point, continual
boiling takes place. It is usual, therefore, to maintain a small positive
pressure in these vessels. This reduces boiling, enables the boil off gases to
be extracted for re-liquefication or flaring, and finally avoids the risk of a
vacuum forming in the tank. In pressure vessel terms this pressure is
low,[27,29] viz. 140 mbar, but its presence influences the type of roof selected.
This is usually a self-supporting structure spanning the tank of domed
shape, the design of which is given earlier in Section 2.2.1. In addition to the
pressure the roof must be capable of sustaining a 6 mbar internal vacuum.

2.4.3. Suspended Roofs—Double-Walled Tanks

Suspended roofs or decks are now widely adopted in these tanks. They are
suspended from the outer roof by means of stainless steel wire ropes or
rods, and used to support the insulation material. By this means the outer

roof remains at approximately ambient temperatures. The deck has been constructed from plywood, aluminium and stainless steel with equal success (see Fig. 5 for the location of the deck).

2.4.4. Choice of Tank Materials
In view of the very low design temperatures in these tanks, special consideration must be given to the choice of materials for the inner tanks. When a double tank is used the requirements for the outer tank are those of normal atmospheric temperatures. In the case of the inner tank the austenitic stainless steels, 9 % nickel ferritic steel and certain aluminium alloys all exhibit excellent ductility at cryogenic temperatures. For the low temperature range (0– – 50 °C) where a single tank may be used, the carbon manganese ferritic steels with low temperature properties can be used. Caution must always be exercised and reference made to the codes.[27–9]

2.4.5. Pipework Connection to the Tank
In most installations as many as ten pipework connections have to be made to the vessel. They are used to fill and discharge the vessel, to exit the boil-off, etc. In recent designs these enter/exit nozzles are located on the roof and penetrate to below the suspended deck. The only line to penetrate the inner tank is the overflow line which is located at the top of the inner tank and routed down the interspace to exit through the outer tank wall near the base. By this means the load-carrying shell is a continuously stressed structure without the stress concentrations, and therefore the inherent uncertainties, of the nozzles.

2.4.6. Outer Bund
A high concrete bund is located as shown in Fig. 5. It is designed to withstand the full LNG contents plus 10 % in the event of a collapse of the metal tank. It has no pipework or cable penetration.

Further developments in this field are taking place involving the use of outer concrete tanks and also the use of internal insulation, i.e. on the inside surface of the inner tank.

3. HORIZONTAL CYLINDRICAL VESSELS WITH DISHED OR FORMED ENDS

These vessels are produced in a wide range of sizes for the storage of gaseous and liquid products. They consist essentially of a circular cross-section cylinder, generally of constant wall thickness, with torispherical or hemispherical ends (see Fig. 6). It is recommended that the vessels be supported at two cross-sections only. This avoids the problems likely in a multi-support system when relative settlement of one or more supports occurs or the variation in the straightness, local roundness and the stiffness of the vessel takes place. These can change the distribution of the reactions on the various supports.

In some cases, however, the multi-support system may be economically more attractive, in which case a complete ring and leg support is preferable (see Fig. 6(d)). The design of the support which will be most suitable for these vessels has received some attention in recent years and mention will be made of these studies. Vessels of this type are used for:

1. the storage of gases in high pressure receivers;
2. liquefied gas storage in double-walled vessels in refrigerated conditions at atmospheric pressure, or a semi-refrigerated condition with pressure,
3. liquid storage at low values of internal pressure.

Discussion follows on these vessels.

3.1. The Storage of Gases in High Pressure Receivers
The requirement of any national or regional gas supply company, which in the UK is the British Gas Corporation, is to provide gas at the correct pressure, calorific value, etc., in the quantities required by the domestic and industrial customer. Because of the variation in gas demand throughout the day, it is necessary to incorporate diurnal storage into the system, i.e. make available some means of storing excess gas during the night and releasing it during peak periods during the day. To cope with this, the familiar low pressure spirally-guided and column-guided gasholders were introduced 150 years ago. However, in recent years other solutions to this problem have been considered. These are discussed in detail by Clarke et al.,[33] with particular reference to the UK scene. One such solution which is widely used is the provision of high pressure receivers of approximately 3·7 m in diameter. When used without compressors they operate between 3 and 25 bar and with compressors between 15 and 70 bar. The advantage of these high pressure receivers compared with the low pressure gasholders, is that they are less conspicuous and provide a large concentration of storage in a given location. On the other hand, they are invariably located in areas of high density of population and are therefore subject to very stringent and costly safety and integrity checks.

An alternative to the high pressure vessels is to store the gas in buried

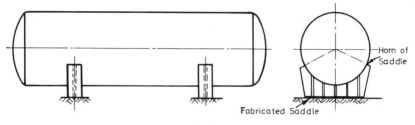

(a)   SADDLE SUPPORTS

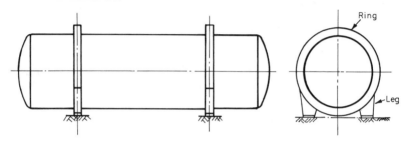

(b)   RING AND LEG SUPPORTS

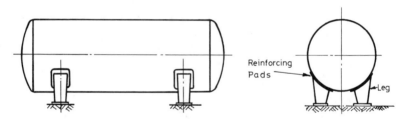

(c)   LEG SUPPORTS FOR SMALL VESSELS

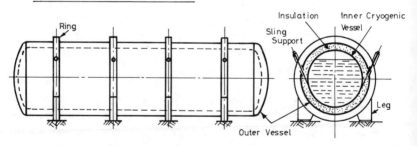

(d)   DOUBLE WALLED CRYOGENIC VESSEL

FIG. 6.   Supports for horizontal cylindrical vessels.

high pressure mains laid solely for this purpose in a pipe array. Such schemes can be located in isolated places and after installation the area can be landscaped and the land above the array farmed as pasture land. A discussion of such schemes with comparisons is given by Sykes and Brown.[34]

An extension of this idea is to use the actual transmission system to provide storage. This is known as 'line pack'. A larger diameter main is installed than is actually needed for transmission purposes and the initial pressure is increased.[33,34] It could well be that 'line pack' will provide the main diurnal storage of the future for British Gas. Nevertheless, a large number of high pressure receivers are still in use in the UK and will continue to provide localised storage. Furthermore such vessels albeit somewhat smaller are widely used in process plant and a design procedure is therefore important.

For vessels of this type the governing design criterion is invariably the high internal pressure which, when applied, results in a relatively thick-walled vessel of R/t ratio ranging from 30 to 100. Because of this it is not necessary to place the supports near the end of the vessel, which would provide diametral stiffness, nor is it necessary to provide a complete ring girder at the support location as shown in Fig. 6(b). Most vessels are thus completely unstiffened in the support region and supported on saddles which only extend part of the way round the circumference (see Fig. 6(a)). It has been found that such vessels are capable of sustaining the high circumferential bending moments which occur at the highest point of the saddle support region, known as the saddle horn (see Fig. 6(a)). Because of this it is possible to use long vessels of 70–100 m in length and locate the saddle support several diameters from the stiffening effect of the ends of the vessel. These vessels are often referred to as 'bullets', an expressive name in view of the potentially lethal explosive energy they contain. Locating the supports away from the ends has the beneficial effect of reducing the high bending moment which occurs at the centre of the vessel.

3.1.1. The Type of Support

In general the supports for the large gas receivers are of fabricated construction in the form of a varying depth 'I' beam or 'T' section. In view of the overall weight of the vessel, plus the extra water weight during the hydraulic test, it is necessary to design a support which embraces the vessel over an included angle which can be of any value between 120 and 150°. The central web is strengthened at regular intervals with stiffeners at right angles and on either side of the web. Although loose saddles are occasionally used, for this application the top saddle plate is welded to the vessel. Sometimes an extra wider plate is used between the vessel and the saddle plate—known as a doubler plate. The reason for using this plate is probably that the vessel and doubler plate undergo heat treatment in the absence of the saddle which is welded on later. The base saddle plate is located on the concrete foundation plinth. One saddle is fixed to the foundation with holding down bolts while the other saddle is free to move axially with the use of a roller, rubber inserts or PTFE bearing pads. This incorporates an element of flexibility and avoids induced stresses due to overall temperature variation, etc.

Some of the saddles used may be considered very rigid compared with the vessel. Others are rigid only over the lower support region and consist of a varying section cantilever from the base up to the saddle horn. This latter type may be considered flexible. It has been found experimentally by Tooth et al.[35-8] that the flexibility of the saddle has a marked effect on the stresses set up in the vessel. For example, when a rigid saddle is used very high vessel/saddle interface forces are set up at the horns.[37] These high values of interface forces cause large circumferential stresses in the vessel at the horn. This is reduced when a more flexible support is employed. It is further reduced when a loose saddle is used and a mismatch of vessel and saddle radii occurs so that there is a small initial gap at the horn (see reference 37 and Greatorex and Tooth[39]).

3.1.2. Analysis and Design
The stresses and displacements in a cylindrical vessel which is ideally circular in cross-section, can be determined from shell analysis provided that the applied forces on the vessel are known. This is so for the torispherical or hemispherical ends used on these cylinders where the approach of BS 5500[5] or ASME Section VIII[57] can be followed. The difficulty with the saddle support problem is that although the self-weight, liquid loading and internal pressure are all known, the distribution of the interface reaction forces in the saddle area are not known. Since the vessel stresses are maximum in this region it is clearly of profound importance that these interface forces be determined for the particular supports being used.

The first comprehensive treatment of this problem was presented by Zick[40] in 1951. Such was the insight of Zick that this approach still forms the basis of the relevant section in BS 5500[5] which is used by most designers in the UK. The approach was semi-empirical in that beam and ring analyses were modified so that the mathematical model used for the vessel predicted stress values which agreed with the experimental results he had available. The basis of these formulations as they are used in BS 5500[5] are given in detail by Tooth.[41]

More exact analyses have been provided by other workers, notably Krŭpka,[42,43] who developed a semi-bending theory, and by Tooth *et al.*[35–37,44,45] who used the complete shell equations. The approach of the latter author is as follows:

1. to represent the known forces—self weight, liquid fill and pressure—as a double Fourier series, giving the radial pressure p_r at a general point (x, ϕ) in the cylinder of length, L

$$p_r = \sum_{n=0}^{\infty} \sum_{m=1}^{\infty} p_{n,m} \cos n\phi \sin m\pi x/L$$

   where $p_{n,m}$ corresponds to the applied loading terms; using the shell equations it is possible to determine the radial and tangential displacements in the saddle region, due to these forces, assuming the vessel is supported at the ends;

2. to divide the saddle support reaction into a series of discrete but as yet unknown radial and tangential forces uniformly distributed over discrete areas which extend across the width of the saddle;

3. to determine the vessel and saddle displacements due to unit saddle forces; then, using the influence area method, to formulate the total radial and tangential displacement, due to the forces, at any point within the saddle area in terms of the unknown interface forces;

4. to use overall equilibrium and saddle compatibility to determine the unknown interface forces;

5. to apply these interface forces to the vessel to determine the stresses, etc., in the vessel.

The details of the numerical techniques and difficulties are given in the above references. In addition to this theoretical work, Tooth *et al.* have carried out experimental work using a series of different vessels. These have included a series of 0·45 m diameter steel pipes,[46] aluminium[44] and perspex[37] cylinders, three 7·32 m long steel vessels of 0·91 m and 1·83 m diameter[36,38] and finally a 3·7 m diameter, 54·9 m long steel gas receiver.[38] On the basis of this experimental and theoretical work it is possible to draw the following conclusions:

1. that the hydraulic test, which is a prior requirement for service, constitutes an exacting case often more severe than the operating

gas-filled condition; in this case the maximum stress occurs in the circumferential direction at the saddle horn; the stress gradient is very high and the stress rapidly dies away;)

2. that for the liquid-filled case the measured stress values for the stiffened 'rigid' saddle can be as much as twice that predicted by BS 5500;[5]

3. that when the saddle is flexible there is good agreement between BS 5500 and the hydraulic tests;

4. that the tests have confirmed that in the liquid-filled case the analytical approach (references 45, etc.,) provides a method of analysis capable of predicting the maximum stress in the vessel with a high degree of confidence;

5. that the load to cause first yield is much smaller than the collapse load and thus there exists a considerable reserve strength even when a vessel is stressed up to yield in the support region;[46]

6. that when the vessels are subject to internal pressure the resulting stress values are highly dependent upon the vessel shape; of particular importance are the small deviations from the ideal circular shape which occur at the horn of the saddle due to the saddle horn weld; the local stresses (see reference 38) are very sensitive to these deviations which are not considered in the theoretical analysis.[45]

At this present time the theoretical results[45] have not been prepared for inclusion into BS 5500[5] and thus the designer is advised to use the code with some caution bearing in mind the above conclusions.

It is hoped in the near future to make available a British Standard dealing with the dimensional details of the saddle supports (see reference 47). Although this is still based on the present concepts of BS 5500[5] the proposed code will give helpful details of saddle designs.

It is appreciated that small gas receivers storing LNG or liquid oxygen at pressure, without refrigeration, may not require the complete saddle support but can be supported on legs (as in Fig. 6(c)). The pad support in this case can be designed assuming a uniformly distributed loading (see reference 5).

3.2. Liquefied Gas Storage in Double-Walled Vessels

As discussed in the earlier Section 2.4 liquid natural gas (LNG) and liquid petroleum gas (LPG) are generally stored in bulk in refrigerated conditions in large vertical cylindrical tanks. For smaller quantities of liquid gas, or when partial refrigeration can be used with pressure, horizontal cylinders and spheres are employed. The cylinder has been found useful for the storage of liquefied oxygen and nitrogen in refrigerated containers at $-183\,°C$ and $-196\,°C$, respectively, and at essentially atmospheric pressure. Double-walled cylindrical vessels are provided with a vacuum/perlite powder insulation infill. In general vacuum insulation, the effect of which is to substantially decrease the thermal conductivity of the inter-cylinder space, is only used for small laboratory installations, or in equipment required to operate substantially below $-200\,°C$.[32] However, when it is combined with the loose fill perlite it is possible to achieve adequate insulation with a lower vacuum condition.

The outer vessel is of carbon steel while the inner vessel must be manufactured from an austenitic stainless or 9 % nickel ferritic steel capable of exhibiting ductility at cryogenic temperatures (see Section 2.4.4). The outer vessel must be designed for vacuum conditions which invariably require that stiffening rings are placed round the vessel at regular intervals along the length (see BS 5500[5]).

These rings can be used to provide an integral base support or 'ring and leg'—Figs. 6(b) and 6(d)—for the whole installation. One method of supporting the inner vessel is by means of adjustable stainless steel straps anchored to the outer vessel at the ring support locations as shown in Fig. 6(d), and by Blenkin.[48] In this case the ring support has to be designed for the additional forces applied by the straps and to avoid the 'hot spots' which cause high local thermal stresses.

When more conventional saddles are used it is necessary to insulate the saddle from the cryogenic vessel. The saddle is thus loose and therefore only carries radial interface forces. In this case it is found, from the theoretical work of Tooth and Duthie,[49] that the peak circumferential stress values are of similar magnitude to those of the welded saddle and therefore higher than those predicted from BS 5500.[5]

3.3. Liquid Storage at Low Values of Internal Pressure

When vessels are used for liquid storage and the internal pressure is small, it is possible to use a relatively thin-walled vessel. The incentive to do this is greater when the more expensive austenitic stainless steels are required as in the food, beverage and pharmaceutical industries. The result is that vessels with R/t values ranging from 100 to 625 are employed. In these cases the wall thickness is in general not adequate to carry the high circumferential moment referred to earlier in Section 3.1 and therefore either

1. the position of the support is within say $R/2$ of the vessel end (or head) so that the vessel benefits from the rigidity of the end; or
2. a 'ring and leg' support is used in the saddle region (as in Fig. 6(b)).

In both of these cases, since R/t is high, attention has to be given to the effect of the high compressive stresses in the longitudinal direction. When the vessel is completely full the maximum value of the compressive force occurs at the top dead centre, or zenith, at the vessel centre. However, for a partially-filled vessel the compressive stress at sections slightly above the height of the fluid can, in certain cases, be as great as that occurring in the full vessel at the zenith. Buckling is thus a possibility which could occur in both partially and completely full vessels.

In both cases it can have catastrophic consequences. This fact is heightened when it is appreciated that actual storage vessels cannot be manufactured without imperfections which, if situated in the critical areas, could well initiate a buckling collapse of the vessel. For this reason the design codes incorporate large factors of safety and lay down limitations on the allowable lack of circularity of the vessel as manufactured, in the hope that a safe vessel will be provided. In addition to the cylindrical shell it is also necessary to design the dished ends to prevent buckling occurring during the internal pressure loading. The shell and ends will therefore be considered in the sections to follow.

3.3.1. Design of the Cylindrical Section

Since the internal pressure is relatively low, of the order of 2·5 bar, and the vessels can be made relatively thin (R/t between 100 and 625), the design requirement is to carry the liquid loading without buckling occurring in the vessel. Like the gas receivers the aim is to position the supports to equalise the bending moment at the vessel centre and support. This is done by assuming the vessel behaves like a simple beam. For this case the ring and leg support is preferred. In other cases the vessel is placed on concrete saddles but here it is necessary to locate the saddles near the ends, within $R/2$. In this case some flexibility of the support is introduced by placing a layer of thick rubber between the saddle and the wear-plate welded to the vessel.

It is found in these vessels at the higher end of the R/t range that large displacements (i.e. greater than the vessel thickness) occur during filling, particularly at the central section. As the vessel is progressively filled, so the cross-section 'rounds up' due to the hydrostatic pressure of the liquid on the vessel walls. When completely full the displacements are in fact relatively small. In such cases the partially full vessel is more likely to buckle, due to high longitudinal stress, than when the vessel is completely full. The likelihood of buckling is further removed during over-pressurisation. On the other hand a small vacuum developing, say during emptying, can dramatically influence the onset of buckling and should be avoided with adequate venting.

A theoretical analysis of the vessel full of fluid supported at the ends was carried out by Stern[50] and later by Tooth and Fernandez[51] using the Galerkin method. The latter showed that the critical longitudinal stress obtained for this case corresponded closely to that of the simply supported cylinder under uniform longitudinal compression, viz.

$$\sigma_c = Et/\sqrt{3(1 - v^2)R} \qquad (10)$$

Putting $v = 0.3$, gives

$$\sigma_c = 0.605 Et/R \qquad (36)$$

A similar theoretical approach was used for the partially filled vessel.[52] The results for this show that the short cylinders of $L/R = 5$ (where $L = $ length of vessel) for values of R/t between 100 and 600, although safe when full, are likely to buckle during filling when the vessel is approximately 73 % full. The same conclusion was reached for $L/R = 10$ for R/t between 300 and 600 cases. The vessels 'at risk' have an $L/R \leq 10$ and therefore particular attention should be exercised in these cases at the design stage. For the longer vessels with $L/R \geq 15$ if buckling does occur it will occur when the vessel is full.

In order to cover these eventualities, together with vessel imperfection, an allowable stress of

$$\sigma_c = 0.060 Et/R \qquad (37)$$

has been suggested, i.e. a factor of 10 on eqn (36). An alternative to this is that proposed in BS 5500[5] and outlined above in Section 2.1.2.2. The stress obtained from either eqn (37) or reference 5 is to be a limitation on the stress calculated when the vessel is full—a procedure which will in fact reduce the above factor of safety for the shorter vessels. The above proposals are, however, based on successful operating experience of many years and therefore give confidence to the approach. In addition to the longitudinal stress, the tangential shear stress should be determined using BS 5500[5] and maintained within the allowable values.

In vessels of this type it is acceptable to use a variation of shell thickness along the length of the vessel, provided that a constant thickness plate of length equal to $R/2$ be used on either side of the support.

3.3.2. Buckling of the Dished End
Vessels of this type are closed at their ends by shells of revolution which are

ellipsoidal or torispherical in shape. They are invariably thin with diameter/ thickness ratios in the range $500 < D/t < 1500$ and can fail under internal pressure by elastic or elastic–plastic buckling. This is caused by the high compressive circumferential forces which are generated in the knuckle region of the dished ends. The buckles form at right angles to these forces and in a progressive manner as the pressure is increased. The behaviour does not appear to be influenced by the lack of symmetry created by the first and consequent buckles since catastrophic failure of the end does not occur. In this respect it may be considered a 'safe' failure. It is as if the end was attempting to deform into a hemisphere. Nevertheless, the presence of the buckles, whether they be elastic and disappear when the pressure is reduced or plastic and remain, should be avoided.

Unfortunately, an adequate and comprehensive set of design rules does not exist for these very thin ends due no doubt to the complexity of the problem. However, in recent years a great deal of experimental work has been carried out on ends of this type and some optimism is being expressed. To this end recent papers by Galletly[53,59] show the value of a large-deflection, elastic–plastic shell bending theory in providing critical pressure values for ideally shaped ends. The results of these are presented for different heads in a series of simple equations. The hope is that these can be extended to cover the as-manufactured vessels, a series of which have been tested by Stanley.[60,61]

4. SPHERICAL OR MODIFIED SPHERICAL VESSELS

In those cases where pressure is a necessary requirement for storage, and is the predominant loading, the spherical vessel has certain advantages.

In the first place the membrane stress, at locations removed from the support, is constant in all directions and lower, by a factor of two, than the maximum membrane stress in the cylinder. It is thus possible to use a thinner wall than in the cylindrical shell. Furthermore the ratio of tank surface to contained volume is smaller for the spherical vessel than for cylindrical vessels—i.e. one can store more liquid for a given surface area of plate material which, when high cost vessel materials are required, can outweigh the high cost of manufacturing the sphere. The spherical shape is thus particularly suitable for storage under pressure. It is less so for the storage of a large quantity of fluid under modest pressures, i.e. where the weight of the fluid constitutes the predominant loading. For this case a modified spherical vessel called a spheroid has been developed. The intention here is to produce a vessel which would have equal stresses at all points (i.e. like the sphere under pressure) and therefore a minimum thickness but support the self weight of the fluid in a membrane manner. The shape chosen is that of a drop of mercury resting on a flat surface. Here the surface tension forms a confining skin comparable with the steel shell of the storage tank and is ideal because the stress in the skin of the drop is equal in all directions. These two types of vessels will be discussed in detail.

4.1. The Spherical Vessel
Spherical vessels are used for the storage of volatile liquids and gases under pressure in the small and medium scale bulk storage. They are located at loading terminals and refineries or may be used as storage within the process plant. The vessels provide a facility for the storage of liquid gas either

1. under partially refrigerated conditions (i.e. higher than the boiling point at standard temperature and pressure) and with internal pressure existing; or
2. under total refrigerated conditions to a temperature equal to or below the boiling point of the liquid gas and at essentially atmospheric pressure.

4.1.1. Partially Refrigerated Vessels
It is found that the partial refrigeration plus pressure condition is quite often more economical than full refrigeration for medium capacity storage. The design pressure and temperature can be chosen for certain gases, such that the minimum storage temperature is above ambient and the existence of the pressure within the vessel is a useful plus factor in the process plant. The value of this pressure, of course, varies with the gas concerned and the operating temperature. It builds up in the vessel to an equilibrium value where no further boiling takes place—for example, if a temperature of 45 °C is selected the pressures for propane and butane are approximately 14 bar and 5 bar, respectively. For such an installation, therefore, cooling would only be necessary (in the UK) in the summer months. Figure 7(a) shows a series of such vessels.

A further advantage of the partial refrigeration is that sometimes less expensive materials can be used since the design temperature is higher. In fact selection of appropriate working pressure and refrigeration plant often permits a wide range of liquefied gases to be stored in the same vessel—thus making prime use of plant facilities. In general, vessels used in partial

FIG. 7.   Spherical vessels. (Courtesy, Whessoe Heavy Eng., Ltd.) (a) Propane and butane storage (16·0–11·1 m diameter); under pressure at ambient temperatures.

FIG. 7.—*contd.*   (b) Ammonia storage (21·3 m diameter); single-walled, partial refrigeration, external insulation, internal pressure = 4·1 bar.

FIG. 7.—*contd.* (c) Crude oil storage (23·9 m diameter); 45 000 barrels (7160 m³), 0·81 m pipeline.

FIG. 7.—*contd.* (d) Ethylene storage (18·9 m diameter); double-walled, cryogenic temperature = −104 °C, small internal pressure = 0·7 bar.

refrigeration systems are single-walled with external insulation as shown in Fig. 7(b).

4.1.2. The Oil and Gas Industry

When employed in the oil industry spherical vessels are used for the storage of the more volatile grades of natural gasoline. They are capable of containing the liquid and its vapour so that there are no filling or breathing losses (see Section 2.2 for discussion of this point). The value of the internal pressure developed is, of course, small compared with the gas liquefaction systems. However the fact that spherical storage is employed does mean that the overall pressure of the total system can be maintained at a high level if required (see Fig. 7(c)). Spherical vessels are also used for gas storage where relatively small quantities of gas are required, for example, for gas enrichment purposes. In this case the gas is simply stored at high pressure at ambient temperatures.

4.1.3. Cryogenic Storage

When cryogenic storage is necessary then double walled vessels are used with an insulation infill of loose perlite between the walls. The problem of consolidation of the insulation, referred to in Section 2.4.1, must be considered in these cases. Various methods of supporting the inner vessel have been proposed:

(a)   by using adjustable slings (similar to the cylinders shown in Fig. 6(d)) from a ring on the outer vessel;

(b)   from rollers located on vertical brackets fixed at the equator of the inner vessel and situated immediately above the outer vessel columns; the load of the inner vessel and contents is thus carried by the columns which extend through the outer vessel;

(c)   from its own set of column supports built to pass through the outer column supports to the foundations; the insulation material is placed in the space between the inner and outer columns; Fig. 7(d) is an example of this system; in this case during construction (i.e. at ambient temperatures) the inner columns are set radially inwards, off the vertical, by a distance corresponding to the radial contraction of the vessel during refrigeration; thus, at the operating temperature, the inner columns are exactly vertical.

4.1.4. Design Considerations

The sphere should be designed as a pressure vessel using the allowable stresses, etc., of BS 5500[5] or ASME Section VIII.[57] The vessels are supported by cylindrical columns attached to the shell at the equator in such a way that the centre line of the column is directly in line with the mid-surface of the vessel wall. By this means the support force resultant is tangential to the vessel mid-surface and bending forces at the junction are reduced to a minimum (see Fig. 7). The forces in the columns arise from the total vessel weight, i.e. vessel material, insulation and liquid contents, plus the forces caused by the wind. The total horizontal wind force on the vessel can be obtained as detailed earlier in Section 2.1.4. The way in which this is carried by the columns is a function of the stiffness of the various structural elements. However a satisfactory design approach to this can be obtained in a simple manner as presented below.

4.1.4.1. Determination of the axial force in the columns. The moment of the wind load about the base of the columns tends to overturn the vessel about an axis normal to the wind direction, while the total weight of the vessel tends to hold it down on its foundations.

The highest compressive stresses in the columns occur on the leeward side, when the vessel is full. The highest tensile stresses are set up on the windward side when the vessel is empty. The compressive stresses are thus the determining factor. The total compressive force in the most remote column on the leeward side is,

$$P = \frac{4Fh}{nd} + \frac{W}{n} \qquad (38)$$

where F = wind force (see eqn (9)), h = vertical distance of resultant wind force to foundation, n = number of columns, d = diameter of column centres (ideally vessel diameter), and W = total weight of vessel, insulation + liquid.

4.1.4.2. Determination of the horizontal force in the columns. In this case the total horizontal force is divided between the columns. As a design rule the following is often used:

No column takes more that $2F/n$ (39)

In order to provide additional restraint and strength for the columns cross bracing is often used as shown in Fig. 7. For the most part the bracing is from the base of the column to the lowest point of the vessel/column junction (as in Fig. 7(a)). However, for particularly tall columns, two sets of bracing are employed as in Fig. 7(c). The bracing is designed to act only in tension and to carry the horizontal force on the column (eqn (39)).

4.2. Modified Spherical Vessels—Spheroids and Hemispheroids

4.2.1. *The Spheroidal Vessel*

As indicated above the spherical shape has been modified to take advantage of the geometric shape of the droplet where the surface tension forms a confining skin comparable with the steel shell of the tank. In point of fact the ideal shape has been further modified from that originally conceived and an outside base ring girder and a series of brackets have been added to support the overhanging load on the shell when the pressure on the inside is less than the maximum design pressure as shown in Fig. 8.

FIG. 8.   Plain spheroidal vessel. (Courtesy, Motherwell Bridge Eng., Ltd.)

A sand cushion of 150 mm thickness is recommended as a base infill, since the sand can be shaped to fit the contour of the tank bottom. Hydrostatic pressure on the base should be prevented by adequate drainage of the site to reduce the height of the water table.

These vessels prevent evaporation losses (see Section 2.2) from volatile liquids by making use of the fact that no loss can occur unless vapour escapes. As with the partially refrigerated liquid gases, boiling continues only until a pressure sufficient to stop it develops inside the vessel. The value of this pressure can be calculated from the maximum and minimum temperatures and the corresponding vapour pressures. The value obtained is entirely independent of the volume of the vapour space. This means that the vessel is just as effective in preventing evaporation loss from standing storage whether nearly empty or full.

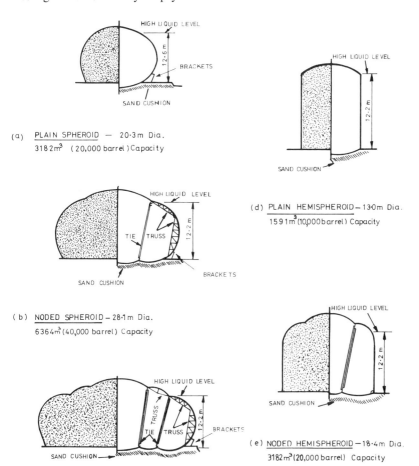

(a) PLAIN SPHEROID — 20·3m Dia.
     3182m³ (20,000 barrel) Capacity

(b) NODED SPHEROID — 28·1m Dia.
     6364m³ (40,000 barrel) Capacity

(c) NODED SPHEROID — 38·9m Dia.
     12729m³ (80,000 barrel) Capacity

(d) PLAIN HEMISPHEROID — 13·0m Dia.
     1591m³ (10,000 barrel) Capacity

(e) NODED HEMISPHEROID — 18·4m Dia.
     3182m³ (20,000 barrel) Capacity

FIG. 9. Spheroidal and hemispheroidal vessels. (Courtesy, Motherwell Bridge Eng., Ltd.)

From the working pressure value and the specific weight of the fluid the vessel can be designed. The analysis is given by Timoshenko[54] and also Flügge[10] who develop the fundamental equations and indicate a procedure for numerical integration. Den Hartog[11] provides a graphical approach. Other helpful details are given in API 620.[28] The smooth spheroidal vessel becomes too costly for storage capacities greater than 40 000 barrels (6364 m^3) and a more economical design is obtained by noding the smooth shape (compare Figs. 9(a) and (b)). Nodes are added to the roof and base.

Internal trussing is introduced to support the curved portion of the shell and the nodes in the roof. The structural members connecting the top and bottom node circles help to support the roof when the spheroid is not subjected to an internal pressure and serve as ties when there is internal pressure in the vessel. Additional capacity is achieved by further noding (see Fig. 9(c)). The vessels shown in Figs. 9(a), (b) and (c) are designed to operate at 1 bar for (a) and 0·7 bar for (b) and (c).

4.2.2. The Hemispheroidal Vessel
This type of vessel combines the features of the circular cylindrical vessel, i.e. ease of manufacture of the shell with the spheroid shape. Examples of the plain and noded designs are shown in Figs. 9(d) and (e).

A great many combinations of spheres and parts of spheres can be used for special storage requirements (see Brownell and Young[4]). The concept throughout is to make use of their membrane load-carrying capacity with the acceptance of edge forces and moments occurring only at junction points.

5.   SPECIAL STORAGE VESSELS

In this section brief mention is made of several different storage facilities not discussed previously. Each example raises important points which are of value generally.

5.1. Vertical Cylindrical Vessels on Columns (or Legs)
For access reasons many vertical cylindrical vessels are raised from the foundation. Although the continuous cylindrical skirt welded from the junction of the shell and base would provide a relatively easy solution, it hardly solves the access problem. The solution generally is to provide a series of vertical columns—as for the spherical vessels in Fig. 7—located on the base itself, or in brackets, or a ring, on the vertical shell. In each case the

aim is to avoid the region of the knuckle or torus which connects the vertical shell to the base spherical cap or cone, since for pressure loading the knuckle is highly stressed. Guidance for the location and design of these is given in BS 5500.[5] However, for the smaller vessels used extensively in the food and beverage industries, these procedures are unnecessary and generally the legs are fixed in the shell/knuckle region. As with the spheres of Section 4.1.4, the aim is to line up the centre of the legs with the mid-surface of the vertical shell wall so that the resultant load will be carried by the legs without causing bending in the vessel. The tubular leg is thus cut back and welded to a compensating plate which locally covers the shell and knuckle region. Since the internal pressure is low—usually only the hydrostatic head—the pressure stresses in the knuckle region are negligible. Thus, the imposition of the support stresses in this region should not cause concern.

In those cases where these vessels are located outside, the procedures for calculating the wind-induced loads in the columns given in Section 4.1.4, eqns (38) and (39) should be used.

5.2. 'Space-Saver' Storage

The rectangular tank has been used extensively for the storage of liquids at atmospheric pressure. Such tanks have the advantage of making optimum use of premium floor space within the process plant building. However, it is necessary to extensively stiffen the flat side walls of the tank with vertical members to avoid the excessive bending of the plate caused by the hydrostatic loading. The alternative to the flat-sided tank is the cylindrical vessel where the hydrostatic pressure is balanced by a circumferential membrane stress. No stiffening is required and the resulting wall thickness is comparatively thin. However, such a vessel is uneconomic on floor space. The 'space-saver' is a combination of the best features of the rectangular and cylindrical vessels. It consists of a series of radiused sheets or lobes (usually of stainless steel) welded into the essentially flat roof and base. The lobes are not subjected to the high bending stresses but are stressed, almost entirely, in a membrane manner. Because of this they can be manufactured from thin gauge sheet. External carbon steel columns are located at the corners and at discrete positions along the sides to carry the resultant hydrostatic forces. The 'space-saver' design thus incorporates the inherent strength of the cylindrical vessel with the space-saving features of a rectangular unit.

Figure 10 shows such a vessel in the process of manufacture. This can be achieved at lower cost than the conventional rectangular tank and have the

FIG. 10. 'Space-saver' storage tank under construction. (Courtesy, MacLeod & Miller Eng., Ltd.[58])

further advantage of dimensional stability during loading and improved mixing characteristics due to the baffling effect of the lobes.

5.3. Elevated Water Storage—Experience in Large Tanks

In order to provide a pressure head for liquid (usually water) storage the storage tank is placed on a tall tower. The procedure is common enough for small capacity storage but a recent report by Szyszkowski[55] on the behaviour of a 22 m diameter torispherical storage vessel is of interest because of its size. The torus and spherical radii were 3·5 and 26 m, respectively, and the wall thicknesses were 10 mm for the lower and 6 mm for the upper sections. During the hydraulic test a series of buckles in the meridional direction, i.e. at right angles to the circumferential direction were noted in the lower torus region. They were stable, forming at regular intervals of fill, in fact similar to those discussed in the pressure-loaded torispherical end, Section 3.3.2. One presumes that they occurred in the lower section, despite it being thicker, because of the increased head of water at this point (note the vessel was about 8 m high at the central region). The buckles did not appear to reduce the load-carrying capacity of the vessel; in fact the shell adapted to the loading and to a stable equilibrium condition. Despite this fact, the vessel was structurally strengthened internally in the immediate region where the buckles occurred, so that the circumferential force carried by the shell plate was reduced. The behaviour of this tank should sound a note of caution.

5.4. Silo Design—The Dynamic Factor

In concluding this survey on storage vessels brief mention must be made of the use of silo storage for granular materials. In essence the silo is a vertical cylindrical vessel with a conical hopper supported on a cylindrical skirt. They can vary in size up to 12 m diameter and a height very much greater than twice the diameter and handle materials as diverse as iron ore, coal, heavy density chalk, wheat and maize. The age old problem for the use of silos is the uneven or inadequate discharge of material. The aim therefore is to produce uniform discharge with the free surface of the material descending uniformly down the silo wall. This is known as 'mass flow'. Desirable as this condition is, it has been found recently by McLean[56] that mass flow imposes a much greater loading on the silo walls than the so called 'Janssen static' pressure normally used. To cope with what is essentially a dynamic problem where the material suddenly dislodges from the silo walls, it is suggested that the 'Janssen static' pressure be multiplied by a factor varying between 2 and 2·2.

ACKNOWLEDGEMENTS

The author wishes to acknowledge his indebtedness to his erstwhile colleagues and friends who over the years have been generous enough to share their knowledge and experience with him, particularly those at the British Gas Corporation, MacLeod & Miller, Motherwell Bridge, Whessoe Heavy Engineering and on British Standards Institution committees. Without their help this review could never have been written. However, the author confesses, any errors contained are all his.

REFERENCES

1. BRITISH STANDARDS INSTITUTION, *Vertical steel-welded storage tanks with butt-welded shells for the petroleum industry*, BS 2654, 1973.
2. BRITISH STANDARDS INSTITUTION,|*Code of practide for foundations*, CP 2004, 1972.
3. AMERICAN PETROLEUM INSTITUTE, *Welded steel tanks for oil storage*, API Standard 650, 6th Edn, 1978.
4. BROWNELL, L. E. and YOUNG, E. H., *Process Equipment Design*, Wiley, New York, 1959.
5. BRITISH STANDARDS INSTITUTION, *Unfired fusion-welded pressure vessels*, BS 5500, 1976.
6. BRITISH STANDARDS INSTITUTION, *Code of basic data for the design of buildings*, CP3, Chapt. V 'Loading', Part 2 'Wind loads', 1972.
7. BRITISH STANDARDS INSTITUTION, *Fusion-welded pressure vessels*, BS 1500, Part 1, 1958.
8. BRITISH STANDARDS INSTITUTION, *Fusion-welded pressure vessels*, BS 1515, 1965.
9. COTTON, H. C. and DENHAM, J. B., *Unfired Pressure Vessels, 33rd Mid-year Meeting of the American Petroleum Institute's Division of Refining*, Chicago, Illinois, May 1968.
10. FLÜGGE, W., *Stresses in Shells*, Springer-Verlag, Berlin, 1962.
11. DEN HARTOG, J. P., *Advanced Strength of Materials*, McGraw-Hill, New York, 1952.
12. TAYLOR, G., *PhD Thesis*, University of Strathclyde, 1975.
13. RISH, R. F., *The Engineer*, **216**, 669, October 1963.
14. LITTLE, W., *MSc Thesis*, University of Strathclyde, 1970.
15. HOLOWNIA, B. P., *Symposium on Wind Effects on Buildings and Structures*, Loughborough University of Technology, April 1968.
16. KRAJCINOVIC, D., *J. App Mech.*, 995, December 1970.
17. LANGHAAR, H. L. and MILLER, R. E., *Proc. Symposium on Theory of Shells (in honour of L. H. Donnell)*, Houston, Texas, 1967, p. 404.
18. WANG, Y. and BILLINGTON, D. P., *Proc. ASCE, J. Eng. Mech. Div. EMS*, 1005, October 1974.
19. DE WIT, J., *Oil and Gas International*, **11**(8), 74–80, August 1971.
20. ADAMS, J. H., *Pressure Vessels and Tanks, 40th Midyear Meeting of the American Petroleum Institute's Division of Refining*, Reprint No. 16-75, Chicago, Illinois, May 1975.
21. SAUNDERS, H. E. and WINDENBERG, D. F., *Trans. ASME*, **53**(15), 107, 1931.
22. ROARK, R. J. and YOUNG, W. C., *Formulas for Stress and Strain*, 5th Edn, McGraw Hill, Kogakusha, 1975.
23. MIXSON, J. S. and HERR, R. W., An investigation of the vibration characteristics of pressurised thin-walled circular cylinders partly filled with fluid, *NASA Tech. Report R145*, 1962.
24. ARNOLD, R. N. and WARBURTON, G. B., *Proc. (A) Inst. Mech. Eng.*, **167**, 62–74, 1953.
25. BARRON, R. and VILLANUEVA, C., *Strain*, 95–105, August 1981.

26. BRITISH STANDARDS INSTITUTION, *The use of structural steel in building*, BS 449, Part, 2, 1969.
27. BRITISH STANDARDS INSTITUTION, *Vertical cylindrical welded steel storage tanks for low temperature service: single-wall tanks for temperatures down to $-50°C$*, BS 4741, 1971.
28. AMERICAN PETROLEUM INSTITUTE, *Recommended rules for design and construction of large welded, low pressure storage tanks*, API Standard 620, 6th Edn, 1978.
29. BRITISH STANDARDS INSTITUTION, *Vertical cylindrical welded storage tanks for low-temperature service: double-wall tanks for temperatures down to $-196°C$*, BS 5387, 1976.
30. THWAITES, A. C. and HEARFIELD, L., The current state of the LNG storage art in British Gas, transmitted by UK Government to *United Nations, Seminar and Study Tour on LNG Peak Shaving, 5-9 March 1979*.
31. CARTER, W. P. and ROWLEY, W. A., *Lloyd's Register Technical Assoc.*, Paper No. 6, Session 1978-79.
32. WARDALE, J. K. S., *Insulation*, 148-54, July 1973.
33. CLARKE, D. J., CRIBB, G. S. and WALTERS, W. J., *The Institution of Gas Engineers, Communication 845*, 108th AGM, May 1971.
34. SYKES, B. A. and BROWN, D., *The Institution of Gas Engineers, Communication 973*, November 1975.
35. WILSON, J. D. and TOOTH, A. S., *2nd Int. Conf. on PV Tech*, San Antonio, ASME, 1973, pp. 67-83.
36. DUTHIE, G. and TOOTH, A. S., *3rd Int. Conf. on PV Tech.*, Tokyo, 1977, pp. 25-37.
37. TOOTH, A. S. and WILSON, J. D., *Strain*, 103-11, July 1980.
38. TOOTH, A. S., DUTHIE, G., WHITE, G. C. and CARMICHAEL, J., *I. Mech. E. Conf. C243*, UMIST, May 1981 (to be published in *J. Strain Analysis*).
39. GREATOREX, C. B. and TOOTH, A. S., *I. Mech. E. Conf. C243*, UMIST, May 1981 (to be published in *J. Strain Analysis*).
40. ZICK, L. P., *Int. A. S. Weld*, **30**, 1951.
41. BRITISH STANDARDS INSTITUTION, *Stresses in horizontal cylindrical pressure vessels supported on twin saddles—a derivation of the basic equations and constants used in BS 5500—Appendix G3.3*, P.D. 6497, 1982.
42. KRÙPKA, V., *Symposium, Pipes and Tanks, IASS*, Weimar, DDR, 15-18 May 1968.
43. KRÙPKA, V., *1st Int. Conf. on PV Tech.*, Delft, 1969.
44. FORBES, P. D. and TOOTH, A. S., *Conf. on Recent Advances in Stress Analysis*, Royal Aero Soc., London, March 1968, pp. 47-58.
45. DUTHIE, G., WHITE, G. C. and TOOTH, A. S., *I. Mech. E. Conf. C243*, UMIST, May 1981 (to be published in *J. Strain analysis*).
46. TOOTH, A. S. and JONES, N., *I. Mech. E. Conf. C243*, UMIST, May 1981 (to be published in *J. Strain Analysis*).
47. BRITISH STANDARDS INSTITUTION, *Pressure vessel details (dimensions)—saddle supports for horizontal cylindrical pressure vessels*, BS 5276, Part 2, 1982.
48. BLENKIN, R., *I. Mech. E. Conf. C243*, UMIST, May 1981 (to be published in *J. Strain Analysis*).

49. TOOTH, A. S., DUTHIE, G., *Euromech 136: Local Problems in Plates and Shells*, Warsaw, September 1980 (to be published in *Archives of Mechanics*).
50. STERN, J., (*a*) *Der Bauingenieur*, **44**(8) 283–8, 1969 (*Mintech. Transl. T6572*; *NEL TT 2172*); (*b*) *Doctoral Thesis*, Fakultät für Banwesen de Technischen Hochschüle Hannover, 1969.
51. TOOTH, A. S. and FERNANDEZ, J. A., in: *Stability Problems in Engineering Structures and Components*, T. H. Richards and P. Stanley, Eds., Applied Science Publishers, London, 1979, pp. 315–40.
52. TOOTH, A. S. and MOHAMMAD, K. J., *Euromech 136: Local Problems in Plates and Shells*, Warsaw, September 1980 (to be published in *Archives of Mechanics*).
53. GALLETLY, G. D., *Proc. Inst. Civ. Eng.*, *Part 2*, **67**, 607–26, September 1979.
54. TIMOSHENKO, S. P., *Theory of Plates and Shells*, McGraw-Hill, New York, 1959.
55. SZYSZKOWSKI, W., *Mechanika Teoretyczna I Stosowana*, **4**(16), 557–71, 1978 (in Polish).
56. MCLEAN, R. F., *Proc. Inst. Eng. & Shipbuilders in Scotland*, Paper 1433, January 1981.
57. ASME, *Boiler and Pressure Vessel Code*, Section VIII, Pressure Vessels, Divs. 1 and 2.
58. MACLEOD & MILLER, ENG. LTD, Glasgow, *UK Patent No. 1589736*, 1981.
59. GALLETLY, G. D., *Proc. Inst. Mech. Eng.*, **195**(26), 329–45, 1981.
60. STANLEY, P. and CAMPBELL, T. D., *J. Strain Analysis*, **16**(3), 171–86, 1981.
61. STANLEY, P. and CAMPBELL, T. D., *J. Strain Analysis*, **16**(3), 187–203, 1981.

Chapter 2

THE STABILITY OF UNDERGROUND CYLINDRICAL SHELLS

P. S. BULSON

*Military Vehicles and Engineering Establishment,
Ministry of Defence, Christchurch, UK*

SUMMARY

A number of analytical methods for predicting the critical radial pressure to cause elastic buckling of soil-surrounded thin-walled cylinders are surveyed. These assume that the soil offers elastic support to the cylinder walls.

Ultimate strength calculations are also presented, taking account of compressive stresses in the cylinder walls that exceed the limit of proportionality of the material. When cylinders are buried horizontally under soil cover, which is then subjected to a surface overpressure, the ratio between the depth of cover and the cylinder diameter is shown to be an important parameter and its effect on collapse strength is examined.

The analysis is linked with a range of experimental results to produce simple rules used in the design of thin-walled culverts, water pipes, and underground structures subjected to static surface pressure.

1. INTRODUCTION

Thin-walled underground structures are becoming more extensively used, particularly in the form of cylindrical shells. Thin-walled culverts made from corrugated or flat metal sheet have been used for water supply and drainage for half a century, particularly in the USA. More recently, road tunnels and underpasses have been constructed from thin-walled tubular construction.

53

These cylindrical structures can prematurely fail by buckling of the side walls, and a good deal of analytical and experimental research has been conducted to enable design checks on stability to be made. In the future, the use of cylindrical modules under the seabed for geophysical and mineral exploration may develop rapidly, and design checks will be required here too.

The purpose of this chapter is to summarise the main findings of the analysis and test programmes, particularly in relation to the interplay between the flexible structures and their supporting soil environments. This will indicate the parameters that need consideration, particularly in judging the effect of stability on ultimate strength. There are two distinct types of loading action: the first when the pressures on the buried cylinders are due mainly to the weight of a superimposed soil backfill, a typical condition for culverts buried under deep embankments, the second when the major load is a static pressure applied to the surface of the soil. It is not proposed to discuss behaviour under dynamic loads in this article.

2. ELASTIC BUCKLING ANALYSIS

In well-compacted soils, and under depths of cover greater than one radius, there is very little deformation of a cylinder as the load on it is reversed, and the cross-section remains sensibly circular until failure by wall-buckling takes place. If the ratio of radius to thickness is high, buckling can occur in the elastic range of wall compressive stress. At low values, plastic buckling will occur, and at intermediate values there is an interaction between the two collapse modes or states.

Let us first consider a thin-walled circular tube of infinite length in a well-compacted medium that we assume has elastic properties, under uniform radial pressure, p, as shown in Fig. 1. The tangential compressive stress, σ_c, is given by

$$\sigma_c = \frac{pR}{t} \tag{1}$$

where R = tube radius, t = tube thickness.

The limiting value of σ_c is the crushing of the wall in compression due to breakdown of material, and this is only reached at low R/t values, approaching zero. For high values it is necessary to calculate the elastic critical radial pressure, p_{cr}, that causes buckling of the wall.

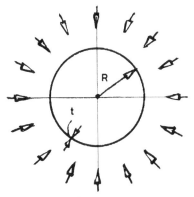

Radial pressure = p

FIG. 1.   Cylindrical tube under radial pressure.

In general, for long tubes, it can be shown that

$$p_{cr} = \frac{(n^2 - 1)}{(1 - v^2)} \frac{EI}{R^3} \tag{2}$$

(see, for example, Timoshenko and Gere[1] (p. 292)) where n = the number of full waves formed around the circumference at buckling, and v = Poisson's ratio. Since $I = t^3/12$, per unit length of tube, eqn (2) can be written

$$p_{cr} = \frac{(n^2 - 1)}{12(1 - v^2)} E \left(\frac{t}{R}\right)^3 \tag{3}$$

A thin-walled tube under uniform radial pressure, but not supported in an elastic medium, e.g. in free air, will buckle into an oval shape with $n = 2$, and in this instance

$$p_{cr} = \frac{3}{(1 - v^2)} \frac{EI}{R^3} \tag{4}$$

If the cylinder is constrained to buckle with large numbers of *full* waves, then n is large compared with unity, and it is sufficiently accurate to write eqn (2) as

$$p_{cr} = \frac{n^2}{(1 - v^2)} \frac{EI}{R^3} \tag{5}$$

In considering the effect of the surrounding soil, we will consider first that the solid provides elastic radial support, as shown in Fig. 2, where the

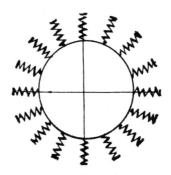

Spring constant=K_z

FIG. 2.   Cylindrical tube with elastic radial support.

spring constant equals K_z (in units of pressure per unit radial deflection). This was analysed by Cheney,[2] who derived the expression

$$p_{cr} = (n^2 - 1) \frac{EI}{R^3} + \frac{K_z R}{(n^2 - 1)} \qquad n \geq 2 \qquad (6)$$

where the buckling mode is given by

$$n = \left[1 + \left(\frac{K_z R^4}{EI} \right)^{1/2} \right]^{1/2} \qquad (7)$$

Substituting in eqn (6) gives the approximate expression

$$p_{cr} = 2 \left[K_z \frac{EI}{R^2} \right]^{1/2} \qquad (8)$$

which underestimates the pressure by less than 10% for $n > 5$, and by less than 1% for $n > 10$. The problem in applying this analysis is to relate the value of K_z to a measurable soil property. If it is assumed that the strain in the surrounding soil is equal to the radial deflection divided by the tube radius, then

$$K_z = \frac{E_s}{R} \qquad (9)$$

where E_s is the soil modulus obtained from triaxial tests at the density and stress applicable to the design conditions. Equation (8) then becomes

$$p_{cr} = 2 \left[E_s \frac{EI}{R^3} \right]^{1/2} \qquad (10)$$

often written as

$$p_{cr} = 2[8E_s E']^{1/2} \qquad (11)$$

where E' is the specific stiffness of the tube, EI/D^3 (and D is the cylinder diameter).

Meyerhof and Baikie[3] devised an expression for p_{cr} at about the same time as Cheney by modifying the theory of buckling of flat plates on an elastic foundation. Using the modulus of subgrade reaction, K_m, with units of pressure per unit radial deflection, they found that

$$p_{cr} = \frac{(n+1)^2 - 1}{1 - v^2} \frac{EI}{R^3} + \frac{(1 - v^2)K_m R}{(n+1)^2 - 1} \qquad (12)$$

which, for larger values of n reduces to

$$p_{cr} = 2\left[\frac{K_m}{(1 - v^2)} \frac{EI}{R^2}\right]^{1/2} \qquad (13)$$

They proposed a conservative relationship

$$K_m = E_s/1{\cdot}5R \qquad (14)$$

so that

$$p_{cr} = 2\left[\frac{E_s}{1{\cdot}5(1 - v^2)} \frac{EI}{R^3}\right]^{1/2} \qquad (15)$$

A further analysis by Luscher and Höeg[4] in 1964 gave, for the higher buckling modes,

$$p_{cr} = 2\left[K_L \frac{EI}{R^3}\right]^{1/2} \qquad (16)$$

and

$$.n = \left[K_L \frac{R^3}{EI}\right]^{1/4} \qquad (17)$$

where K_L is the coefficient of elastic soil reaction (in units of pressure per unit radial strain). If K_L is taken as equal to E_s, eqn (16) becomes

$$p_{cr} = 2\left[E_s \frac{EI}{R^3}\right]^{1/2} \qquad (18)$$

which is identical to eqn (10).

A purer analysis, in which the soil was taken as a true elastic medium,

having a modulus E_s and Poisson's ratio v_s, was carried out in 1964 by
Forrestal and Herrmann.[5] For a long cylindrical shell in which there was no
bond between the shell and the elastic medium (i.e. no shear stresses
present) they found that

$$p_{cr} = \frac{(n^2 - 1)}{(1 - v^2)} \frac{EI}{R^3} + \frac{E_s}{(1 + v_s)(1 - 2v_s)(n + 1)n} \tag{19}$$

at high values of n this becomes

$$p_{cr} = \frac{n^2}{1 - v^2} \frac{EI}{R^3} + \frac{E_s}{(1 + v_s)(1 - 2v_s)n^2} \tag{20}$$

Later Duns[6] used the Donnell shell-stability equation to arrive at a similar
relationship. For the unbonded condition he found that, when $n \geq 7$, there
was an approximate relationship between n, K_L and E_s in the form

$$K_L = \frac{nE_s}{8(1 - v_s^2)} \tag{21}$$

When there was full bonding the value of K_L was increased fourfold. Thus,

$$K_L = \frac{nE_s}{2(1 - v_s^2)} \tag{22}$$

at high values of n, taking $v = v_s$ and neglecting bonding

$$p_{cr} = \frac{2n^2}{1 - v^2} \frac{EI}{R^3} \qquad \text{where } n = \frac{R}{2t} \left[\frac{12E_s}{E} \right]^{1/3} \tag{23}$$

so that, for a given number of full circumferential waves, the value of p_{cr} is
double that given in eqn (5) for buckling in free air. (For a further
discussion of buckling in a soil-surrounded medium, the reader should see
papers by Habib and Phong,[7] and Sonntag.[8])

All the above theories assume that when buckling occurs the elastic
medium stays in contact with the wall of the cylinder. Experiments show
that in many practical conditions this does not take place, and that the half
waves in the cylinder wall which deflect inwards part company with the
medium as they do so. The effect of this was analysed by Chelepati,[9] who
chose an elastic foundation with linear spring characteristics in compression and zero resistance in tension. He found that

$$p_{cr} = (n^2 - 1) \frac{EI}{R^3} + \frac{K_z R}{2(n^2 - 1)} \qquad n \geq 2 \tag{24}$$

and comparing this with eqn (6) showed that the loss of contact introduces a factor of 2 in the denominator of the second part of the expression for p_{cr}. Equation (24) yields the approximate lower bound solution

$$p_{cr} = 2^{1/2} \left[K_z \frac{EI}{R^2} \right]^{1/2} \tag{25}$$

and comparison with eqn (8) shows that the loss of contact reduces p_{cr} in the ratio $1/\sqrt{2}$.

For design purposes, the most popular equations are eqns (10) and (11), which are recommended in a number of design practice documents (for example, by Compston et al.[10] in CIRIA Report 78). Accurate tests by Allgood and Ciani,[11] and by the author Bulson,[12] have suggested that these equations depart from the general trend of test data for deeply buried cylinders, a fact noted by Cheney.[13]

The design equation, eqn (10), can be placed in terms of thin shell properties $I = t^3/12$, and $E \to E/(1 - v^2)$ for thin sheet, as

$$p_{cr} = 0.61 \left[E_s \frac{Et^3}{R^3} \right]^{1/2} \tag{26}$$

Using a theory presented by Cheney gives closer agreement with the trend of results, i.e.

$$p_{cr} = 0.54 E_s^{2/3} E^{1/3} \left(\frac{t}{R} \right) \tag{27}$$

He points out the interesting fact that the two-thirds power has links with the buckling of straight, flat strips on an elastic foundation. In fact, if the critical axial stress in thin sheet supported by an elastic medium is replaced by $p_{cr}(R/t)$,

$$p_{cr} = 0.52 E_s^{2/3} E^{1/3} \left(\frac{t}{R} \right) \tag{28}$$

Thus, for larger values of n, the cylinder behaves like a flat strip supported by an elastic medium.

3.   ULTIMATE STRENGTH

When the calculated value of p_{cr} gives rise to tangential compressive stresses in the wall of the cylinder that exceed the limit of proportionality of the stress/strain relationship for the cylinder material, the ultimate strength, or

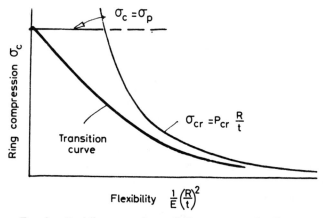

FIG. 3. Buckling curve for radially compressed cylinders.

buckling strength, will fall below p_{cr}. This is analogous to the strength of a simple strut, which falls below the Euler buckling stress as the compressive stress approaches the plastic range.

The value of compressive stress, σ_c, that causes lining failure can be linked to the flexibility of the cylinder by a buckling curve similar in form to well known strut curves. This is illustrated in Fig. 3, which relates σ_c to the flexibility term $1/E(R/t)^2$. At low values of flexibility σ_c approaches the compressive yield stress of the material, σ_p. For materials having no sharply defined yield, σ_p would be the 1 or 2% proof stress. At high values of flexibility, σ_c approaches the critical elastic buckling stress σ_{cr}, which corresponds to p_{cr}. At intermediate values there is a transition between these extremes, and taking a lead from the long established Rankine formula for struts, it can be assumed that

$$\frac{1}{\sigma_c} = \frac{1}{\sigma_p} + \frac{1}{\sigma_{cr}} \tag{29}$$

or

$$\frac{\sigma_c}{\sigma_{cr}} + \frac{\sigma_c}{\sigma_p} = 1 \tag{30}$$

This is sometimes written as

$$\sigma_c = \frac{\sigma_p \sigma_{cr}}{\sigma_p + \sigma_{cr}} = \frac{\sigma_p}{1 + \sigma_p/\sigma_{cr}} \tag{31}$$

The upper limiting value of σ_c is often known as 'ring compression crushing', such as would occur if the cylinder was buried in a medium having an infinite modulus ($E_s = \infty$). The idea of using a transition curve of this type was proposed by Watkins,[14] and the ring compressive strength of non-buckling thin-walled cylindrical conduits has been discussed by White and Layer.[15] This method makes no allowance for the effect of imperfections in the tube, or residual stresses that might be present in large thin-walled tubular structures produced by welding.

4. THE EFFECT OF COVER DEPTH ON COLLAPSE STRENGTH

In most practical designs, the radial pressure around a thin-walled circular conduit is not uniform. If the pressure is due to the dead weight of earth cover there will be a pressure gradient across the vertical diameter of the pipe, and this should be taken into consideration when the gradient is a significant fraction of the mean pressure. If the surface of the backfill is subjected to a high overpressure, then the proximity of the crown of the tube to the surface will have a marked effect on the level of pressure to cause collapse, particularly when the depth of cover to the crown is less than one diameter. The latter case has been the subject of much research and testing, and analyses have been evolved that explain the behaviour of the soil/tube system.

Luscher[16] used the concept of a thin metal tube surrounded by an annular ring of soil, where the radius of the annular ring equals $R + d$, and d is the cover depth over the crown of the tube, as shown in Fig. 4. The width of the annular ring, d, influences the value of K_L in eqn (16).

A method of calculating K_L is to formulate the modulus of resistance of the soil ring (plane strain) to a uniform pressure applied radially inside the cavity formed if the metal tube is removed. Using 'thick cylinder' elastic theory, Luscher found that

$$K_L = \frac{\left(1 - \left(\dfrac{R}{R+d}\right)^2\right) E_s}{(1 + v_s)\left[1 + \left(\dfrac{R}{R+d}\right)^2 (1 - 2v_s)\right]} \tag{32}$$

and this relationship is given in Fig. 5 for $v_s = 0.3$ and 0.5. Note that when $v_s = 0.5$,

$$K_L = \tfrac{2}{3}E_s\left(1 - \left(\frac{R}{R+d}\right)^2\right) \tag{33}$$

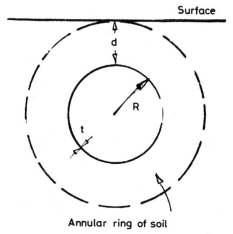

Annular ring of soil

FIG. 4.   Annular ring concept.

Further, when $d = 0$, $K_L = 0$, and when d is large compared to R, $K_L \to \frac{2}{3}E_s$. The value of K_L does not alter much at cover depths greater than $d = 2R$, but there is a lot to be gained in terms of buckling strength if the tube is buried at least $\frac{3}{4}$ diameter deep.

The relationship between collapse pressure and depth of cover has also been examined by the author, Bulson,[17] using the concept of a ring of soil,

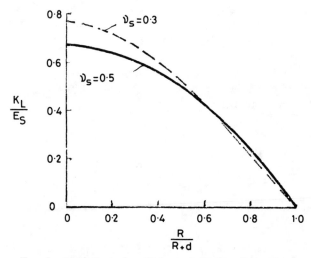

FIG. 5.   Variation of coefficient of elastic soil reaction.

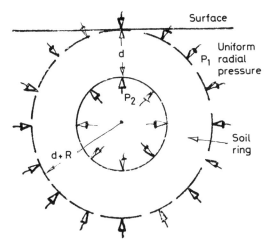

FIG. 6.   Thick elastic cylinder analysis.

radius $(R + d)$, as shown in Fig. 6. By the theory of thick elastic cylinders under conditions of plane stress

$$-\frac{b^2}{b^2 - R^2}(p_1 - p_2) + \frac{p_2 R^2 - p_1 b^2}{b^2 - R^2} = (\sigma_\theta)_R \qquad (34)$$

where $b = R + d$, p_1 = uniform radial pressure on the outside of the soil ring, p_2 = pressure exerted between the tube and the inside of the soil cylinder, $(\sigma_\theta)_R$ = tangential compressive stress at the inner surface of the cylinder. Also,

$$+\frac{b^2}{b^2 - R^2}(p_1 - p_2) + \frac{p_2 R^2 - p_1 b^2}{b^2 - R^2} = (\sigma_R)_R \qquad (35)$$

where $(\sigma_R)_R$ is the radial compressive stress at the inner surface of the soil cylinder. The tangential compressive strain at the inner surface, ε_R, is therefore given by

$$\varepsilon_R = \frac{1}{E_s}\left[(\sigma_\theta)_R - v_s(\sigma_R)_R\right] \qquad (36)$$

where v_s = Poisson's ratio for the soil, E_s = elastic modulus of the soil. Substituting for $(\sigma_\theta)_R$ and $(\sigma_R)_R$ in eqn (36), and remembering that

$$\varepsilon_R = \frac{p_s R}{tE} \qquad (37)$$

where E is the elastic modulus of the material of the tube, leads to

$$\frac{p_1}{p_2} = \frac{1}{2}\left[(1 + v_s) + \frac{E_s R}{Et} + \frac{R^2}{(R+d)^2}\left(1 + v_s - \frac{E_s R}{Et}\right)\right] \quad (38)$$

When d is large compared with R, the last term approaches zero, and further, $E_s R/Et$ is a small number and can be neglected: then, at deep covers $(d \to \infty)$,

$$p_1 = \frac{p_2}{2}(1 + v_s) \quad (39)$$

and at shallow covers $(d \to 0)$,

$$p_1 = p_2 \quad (40)$$

If the thin-walled circular tube is under such a high surface pressure, p_s, and any pressure due to the dead weight of the fill can be neglected, then, by elastic theory,

$$p_1 = 2(1 - v_s)p_s \quad (41)$$

so that, at deep covers, from eqn (39)

$$p_s = \frac{p_2}{4} \cdot \frac{(1 + v_s)}{(1 - v_s)} \quad (42)$$

As the surface pressure is increased, p_2 will eventually reach a value equal to the collapse stress of the tube. If we assume that the R/t ratio for the tube is sufficiently high to ensure buckling in the elastic range, then at failure $p_2 = p_{cr}$. Further, if the value of p_s to produce this condition at deep covers is p_{max}, eqn (42) can be rewritten as

$$p_{max} = \frac{p_{cr}}{4} \cdot \frac{(1 + v_s)}{(1 - v_s)} \quad (43)$$

At intermediate values of cover

$$\frac{p_s}{p_{max}} \propto 1 - \left(\frac{R}{R+d}\right)^2 \quad (44)$$

Using eqn (43) in conjunction with eqn (23), and taking $v_s = 0.3$, gives the useful design relationship

$$p_{max} = 0.12 E_s^{2/3} E^{1/3}\left(\frac{t}{R}\right) \quad (45)$$

which is about half the value for $p_{\max}$ that would be derived from eqn (28), but agrees well with the lower scatterband of a large number of experimental results, as will be shown later.

If tube buckling does not occur in the elastic range, then the value of p_{cr} will need to be modified, using eqn (31), after employing the relationship $\sigma_{cr} = p_{cr}(R/t)$. Thus,

$$p_2 = \sigma_c \, \frac{t}{R}$$

where

$$\sigma_c = \frac{\sigma_p}{1 + \dfrac{\sigma_p t}{p_{cr} R}} \tag{46}$$

5. STATIC TESTS

Tests show that if cylindrical tubes are buried with a very small depth of cover (i.e. $d < R/2$), and the surface is subjected to a static overpressure, collapse usually results from a caving-in of the crown, as shown in Fig. 7(a). At deeper covers, however, collapse in a well-compacted sandy soil usually takes the form of wall-buckling, as discussed in the analysis, and shown in Fig. 7(c). At intermediate depths a combination of the two can take place (Fig. 7(b)). The collapse sequence at deeper covers is shown in Fig. 8. As the pressure is applied the crown of the tube deflects downwards and the sides outwards; the invert has virtually zero deflection. A typical plot of radial deflection during this early loading phase is shown in Fig. 9, and it can be calculated, using the bending theory for rings, that the ratio between total vertical deflection and total horizontal deflection is of the order one would expect if the pressure were uniform around the tube.

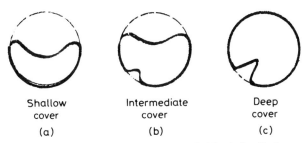

Shallow	Intermediate	Deep
cover	cover	cover
(a)	(b)	(c)

FIG. 7.   Collapse modes of thin-walled buried cylinders.

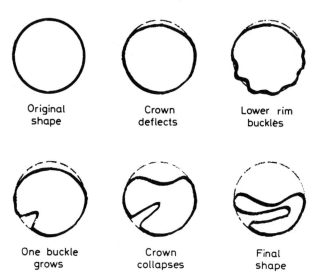

FIG. 8.   Collapse sequence of thin-walled cylinder under deep soil cover.

As the pressure is increased the wall of the tube around the lower half of the rim buckles into a number of elastic half-waves, and one of these buckles grows large. This action causes the diameter of the tube to contract and the crown to collapse inwards as shown in the final drawing of Fig. 8. Nearly all test work in this field has been carried out by model studies,

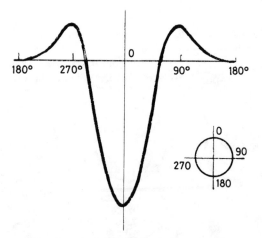

FIG. 9.   Typical radial deflection plot.

usually by burying horizontally-orientated tubes of varying length, diameter and thickness in compacted sand or clay, contained in a closed tank. The tank is usually circular or square in plan, and its bonnet removable. Pressure is applied by compressed air or by water, acting through a diaphragm onto the soil surface. All the available studies up to July 1965 were summarised in a report by Dorris,[18] who also described his own experiments. He compared test results with the theories of Cheney[2] and Forrestal and Herrmann,[5] as shown in Fig. 10.

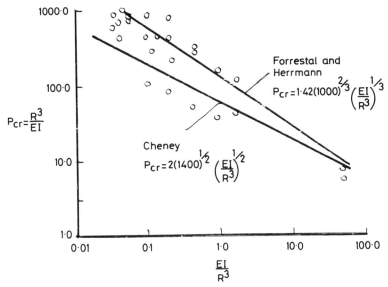

FIG. 10.   Test results compared with theory.

In the late 1960s the author conducted a series of model tests in the rig, shown diagrammatically in Fig. 11, which is described in reference 19. It measured $5\,ft^2$ in plan, $4\,ft$ deep, and was filled with soil in 3 in layers. Each layer was compacted to give a consistent density. When the soil level reached the bottom edge of two inspection ports, in opposite vertical sides of the container, the tube under test was placed horizontally on the soil and in line with the ports. The tube ran the whole length of the rig between the ports and was sliced transversely into five sections to eliminate overall longitudinal bending. The length of the tubes was thought to be sufficiently great to eliminate any arching action in the longitudinal direction between the tube ends.

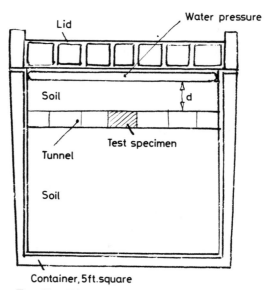

Fig. 11.   Diagrammatic view of static test rig.

Further layers of soil were then laid and compacted until the desired cover was reached. In order to preserve the shape of the tube during this phase it was necessary to strut the tubes with wood sections that were removed after compaction. A pair of soft rubber diaphragms, edged by a steel frame, was then placed on the soil surface, and above this a restraining structure was connected, via links, to the side of the tank. As water was pumped into the space between the diaphragms a uniform pressure was applied over the surface of the soil, measured by a gauge in the water supply

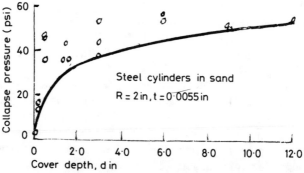

Fig. 12.   Steel cylinders in sand (static surface pressure).

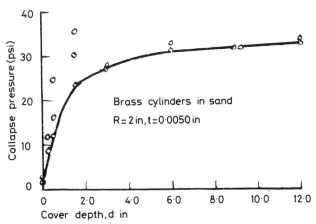

FIG. 13.   Brass cylinders in sand (static surface pressure).

system. The maximum working pressure was 100 psi. By limiting the diameter of the tubes under test to about 4 in, an order of magnitude less than the rig dimensions, it could be assumed that the side walls of the tank had no effect on the action of the specimens under load.

Tests were first carried out for a range of cover depths, two types of tube material (steel and brass) and two types of soil (well-compacted sand and remoulded clay), and the results are shown in Figs. 12–15. The lower scatterband of the results agrees well with the form of eqn (44).

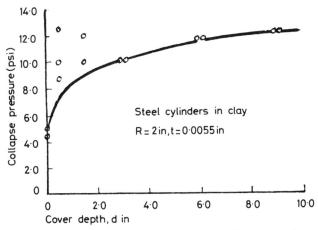

FIG. 14.   Steel cylinders in clay (static surface pressure).

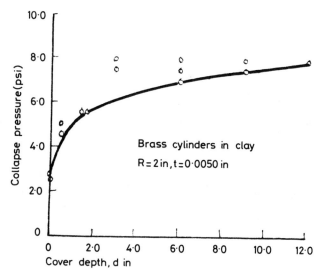

FIG. 15.   Brass cylinders in clay (static surface pressure).

Applying eqn (45) suggests that in the same soil, under similar conditions of compaction, and for tubes having similar radii,

$$\frac{[p_{max}]_{steel}}{[p_{max}]_{brass}} = \frac{[E^{1/3}t]_{steel}}{[E^{1/3}t]_{brass}} \tag{47}$$

which gives a ratio of 1·4 for the tubes under test. From Figs. 12–15 the experimental ratio is also 1·4.

Equation (45) also suggests that, for tubes manufactured from similar material, and having similar radii

$$\frac{[p_{max}]_{sand}}{[p_{max}]_{clay}} = \frac{[E_s^{2/3}t]_{sand}}{[E_s^{2/3}t]_{clay}} \tag{48}$$

which gives a ratio of 4·5 for the tubes under test. From Figs. 12–15 the experimental ratio is 4. The tests therefore strongly supported the analysis leading to eqns (44) and (45).

A further series of experiments was conducted in clay using $\frac{1}{32}$ in thick polyester resin/slate tubes of 4 in diameter, reported in reference 20. The material was chosen deliberately because of its relatively brittle nature, and the influence of this lack of ductility on buckling and ultimate strength was examined in the tests. The results are shown in Fig. 16. Taking

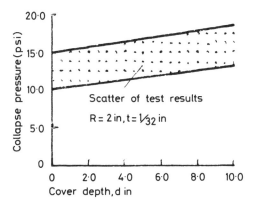

FIG. 16.   Polyester resin/slate tubes in clay.

the modulus of elasticity of the polyester resin/slate (pr/s) as $6 \times 10^5 \, \text{lb in}^{-2}$, and again using eqn (45), gives

$$\frac{[p_{max}]_{steel}}{[p_{max}]_{pr/s}} = \frac{[E^{1/3}t]_{steel}}{[E^{1/3}t]_{pr/s}} = 1 \cdot 5 \qquad (49)$$

As Figs. 14 and 16 indicate, the experimental ratio at deep covers lies in the range 1·0–1·4.

Using the same test rig, an interesting series of experiments was made by Allison[21] to investigate tube behaviour when discontinuities are produced in the soil by the introduction of rigid interfaces. The idea was to give an experimental guide to the minimum dimensions of test tanks to ensure that the walls have no influence on collapse pressure. Brass tubes, 4 in diameter, 0·0050 in thick, were used and heavy, inflexible stone slabs provided the adjustable boundaries.

In the first series of tests the tubes were buried to a cover depth of 9 in, and the slab set horizontally in the soil below the tube. The distance from the tube invert to the slab was varied in successive tests from 0 to 9 in. The results given in Fig. 17 show that the proximity of the invert to the rigid base has no measurable effect on collapse pressure. In the second series the tubes were again buried 9 in deep, and one rigid slab was set 3 in below the invert. Two vertical slabs were then brought progressively nearer to the specimen from each side. The results shown in Fig. 18 indicate that wall proximity begins to have an effect at distances below 1·5 times the diameter from the side of the tube, which compares favourably with the stress

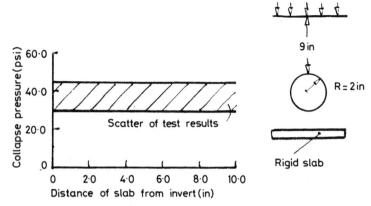

FIG. 17.   The effect of a rigid slab below the invert on collapse pressure.

analysis of holes in a semi-infinite isotropic sheet. Two similar holes placed four diameters apart have no influence on each other.

The tests so far described were concerned with the effect of the surface pressure acting on the soil, when this produces a much more severe loading on the pipe than the deadweight of the backfill. However, a range of practical problems are concerned with pipes or conduits under an external loading of backfill and fluctuations in the ground water table only. The most notable experiments in this field have been conducted under the guidance of Schofield,[22] using a centrifuge to model the effect of gravity stresses in the backfill. English[23] analysed the onset of instability for a buried cylinder with groundwater rising slowly in permeable sand around

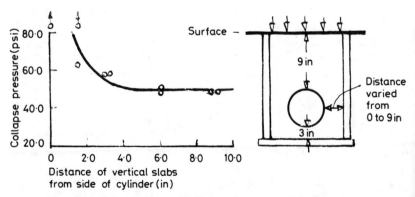

FIG. 18.   The effect of vertical rigid slabs on collapse pressure.

it, and more recently Valsangkar and Britto[24] have conducted centrifuge experiments to investigate the combined action of the self-weight of the backfill and of surface loading. The reader should also see their earlier report (reference 25).

Larger scale tests on flexible pipes have been sponsored by the US Department of the Interior—Bureau of Reclamation, and have been reported on by Howard[26] and Pettibone and Howard.[27] Three equi-spaced diameters were selected, 18, 24 and 30 in (46, 51 and 76 cm), and wall thicknesses of 7, 10 and 14 gauge were used. The pipes were buried in a 6 ft × 7 ft (plan) container, 7 ft deep, so that the longitudinal axes were 4 ft below the soil surface (fully described in reference 27). The surface loads were applied by a large universal testing machine, and were increased until either the pipe failed or until a pressure of 100 psi, the design limit, was reached.

All pipes deflected similarly up to 10 or 15 % diameter change, but thereafter two patterns were followed, depending on pipe flexibility. The stiffer pipes deformed elliptically and failed after the formation of plastic hinges at the ends of the horizontal diameter. The more flexible pipes deformed rectangularly with four plastic hinges forming at 60–70° either side of the vertical diameter. Following this formation, inelastic buckling of the top of the pipe took place.

The relative size of the specimens and the container meant that the results were influenced by the proximity of the container sides to the test pipes. Nevertheless it was felt by the testing team that the relationships established could be finally correlated and refined with the results of field trials on similar specimens.

6. DESIGN RULES FOR STATIC LOADS

6.1. US Corrugated Steel Pipe Institute

The work of Spangler, White and Meyerhof has been used to form design rules for flexible soil-steel pipe structures in the USA. Much of the work has been sponsored by the Corrugated Steel Pipe Institute and by the Armco International Corporation. The latter produce a useful handbook, covering strength, durability, subdrain design, earth control and installation, obtainable by application to the Corporation. The Institute commissioned T. C. McCavour[28] to prepare a design guide in the late 1960s. The first part, dealing with circular pipes under a minimum fill height of one diameter, was issued in May 1966. The second part, issued in December 1968 extended the design procedure to circular pipes under a minimum fill height of one quarter of the diameter.

The guide begins by emphasising the importance of analysing the earth fill and subgrade as well as the strength of the structure, and proposed the following checks:

Classification and evaluation of soil type.
Determination of fill soil properties (above and around the structure).
Determination of subgrade soil properties (below the structure).
Structural analysis of the pipe:

(a)   circumferential ring stress;
(b)   elastic buckling stress;
(c)   axial stress;
(d)   joint strength;
(e)   longitudinal ring stress;
(f)   circumferential deflection.

The last check is very important, because first failures of thin-walled corrugated pipes have been mainly due to excessive circumferential ring deflection. The cause has usually been insufficient compaction of the backfill in critical areas. The economic balance is between the cost of excavation and soil compaction against the cost of the pipe material.

The recommended procedure is to classify the soil type by using the Unified Soil Classification System, and conduct triaxial compression tests on compacted soil samples, maintaining a steady lateral hydrostatic pressure as the axial pressure is increased to failure. From this test the modulus of soil deformation, E_s, is obtained, and, the modulus of soil reaction, K_m, found from eqn (14) or, more accurately, from the following equation

$$K_m = \frac{E_s}{2(1 - v_s^2)R} \tag{50}$$

The relationship between K_m and the depth of cover is then set down on the lines proposed by Luscher[16] (see eqn (33)). Thus,

$$K_m = \frac{E_s}{2(1 - v_s^2)R} \left[1 - \left(\frac{R}{R+d} \right)^2 \right] \tag{51}$$

This follows recommendations by Meyerhof,[29] based on the work of Luscher and Höeg.[4]

Circumferential ring stress is next evaluated, and the work of Allgood,[30] Donnellan[31] and Dorris[18] is used to produce the following formulae:

$$\text{ring thrust, } c = pR \tag{52}$$

where the earth pressure, p, is the sum of that created by the dead weight of the soil and the live loads on the surface, i.e.

$$p = p_D + p_L \tag{53}$$

It is recommended that the proposal of White[32] should be used to find p_D. He suggested that the pressure should be equal to the product of soil density (γ) and the average height of fill above the circular pipe. The average height is approximately $d + 0.225R$, so that

$$p_D = \gamma(d + 0.225R) \tag{54}$$

To determine p_L, Boussinesq's equation is used in the well known three-dimensional form

$$p_L = \frac{3WZ^3}{2\pi r^5} \tag{55}$$

where W is the concentrated surface load, r is the horizontal distance from this load to the point at which the stress is applied to the structure, and Z the vertical distance between this point and the surface.

If A is the cross-sectional area of the corrugated steel conduit per unit length of horizontal projection, the axial stress (σ_c) is given by

$$\sigma_c = \frac{pR}{A} \tag{56}$$

The value of σ_c to produce elastic buckling of the pipewall (σ_{cr}) is found from eqn (13). Thus,

$$\sigma_{cr} = p_{cr}\frac{R}{A} = \frac{2}{A}\left[\frac{K_m}{(1-v^2)}EI\right]^{1/2} \tag{57}$$

where K_m is given by eqn (51). To take account of the inelastic buckling range, eqn (31) is used. McCavour proposed a safety factor of 2 on buckling strength and 3 on the shear or bearing failure of the bolted joint.

Longitudinal ring stress (σ_l), due to longitudinal deflection of the pipe after consolidation and compaction of the soil, is predicted from the analysis of a circular ring supporting a triangular load over the entire length of the pipe. This is given as

$$\sigma_1 = \frac{10RE\,\Delta y}{l^2} \tag{58}$$

where l and Δy are defined in Fig. 19.

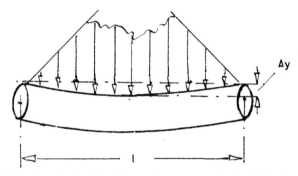

FIG. 19.   Longitudinal bending under a triangular load distribution.

Ring deflection formulae based on the work of Spangler and Meyerhof are recommended. The total vertical diametral shortening, d_c, is given by a suggested formula by Meyerhof[29]

$$d_c = \frac{0 \cdot 167pR^4}{EI + 0 \cdot 061\left[1 - \left(\frac{R}{R+d}\right)^2\right]KR^4} \tag{59}$$

It is normally assumed that if the vertical diameter decreases by about 20 % the pipe is in a state of incipient failure, and that any further vertical load will cause collapse. Measurements of deflections in practical installations have shown that the average vertical deflection of all culverts is about 2·5 % of the diameter. McCavour concludes by giving useful design examples and by discussing materials and design factors of safety.

6.2. Construction Industry Research and Information Association (CIRIA) Recommendations (Report 78)
The CIRIA report, reference 10, contains a critical review of design procedures and examples are given of the proposed methods, which are very similar to those in the US Corrugated Steel Pipe Institute proposals.

The pressure due to dead weight of soil is given as

$$p_D = \gamma(d + 0 \cdot 107D) \tag{60}$$

which is only slightly different to eqn (54), and the pressure due to live loads is given by the Boussinesq equation, eqn (55).
The maximum pressure due to ground water is also specified (p_W) where

$$p_W = \gamma_W h \tag{61}$$

and γ_W = specific weight of water and h = height of water table above the pipe invert. An allowance can be made for a reduction of apparent density of soil resulting from submergence, when the pipes are set in permeable submerged soils. Temporary lowering of the ground water level during installation must be considered. Allowance should also be made for possible 'vacuum' conditions in the pipe. The buckling pressure of the pipe wall is given as eqn (11), and the effect of plastic range buckling is taken into consideration by employing eqn (31), with a safety factor of 3·0.

6.3. Military Vehicles and Engineering Establishment (MVEE) Design Code

The author has recommended design procedures after several years of research at the MVEE, Christchurch, UK, carried out on behalf of the Scientific Advisory Branch of the Home Office, and summarised in reference 12. Much of the research was concerned with underground tubular structures in which the loading due to applied surface pressure was far greater than that due to the dead weight of the soil. In most of the design recommendations, therefore, the latter was neglected.

The surface pressure to cause collapse of a thin-walled buried tube of circular cross-section when the tube is buried under a very large cover depth, is calculated using eqn (45). Thus,

$$p_{max} = 0 \cdot 12 E_s^{2/3} E^{1/3} \left(\frac{t}{R} \right) \tag{62}$$

Note that this formula is based on the theoretical work of Bulson,[17] Duns[6] and Cheney,[13] and on the experimental results of the first-named.

The surface pressure, p_s, to cause collapse at a finite cover depth, d, is found from

$$p_s = p_{max} \left[1 - \left(\frac{R}{R+d} \right)^2 \right] \tag{63}$$

This gives a relationship between p_s and p_{max} that not only agrees with the author's tests on buried tubes, but also agreed closely with the observations of Gill,[33] who plotted all available arching data for plates and cylinders in granular soil.

The allowable surface pressure, p_a, is related to p_s by the equation

$$\eta p_a = p_s \tag{64}$$

where η is the safety factor. For steel and aluminium tubes under static loading the following values are proposed:

Large cylindrical tubes, human habitation, $\eta = 3 \cdot 0$

cylindrical tubes for pipelines or conduits, $\eta = 2 \cdot 0$

For GRP pipes it is proposed that the above values become $3 \cdot 75$ and $2 \cdot 5$, respectively.

When there is no surface pressure present, this code suggests that the soil pressure should be calculated for an imaginary surface, $3R$, above the crown of the cylinder, and that this can be taken as a surface pressure on a pipe having a burial depth of $3R$. If the soil density equals γ,

$$p_s = \gamma(d - 3R) \tag{65}$$

This value can be compared with the value calculated from eqn (63) using $3R$ instead of d, so that

$$p_s = p_{max}\frac{15}{16} \tag{66}$$

7. DISCUSSION

The analytical work reviewed above assumed that soil, whether well-compacted sands or loams or soft clays, could be taken as an elastic medium having properties that varied with depth of cover. Since there are only minor deviations from circularity of the cylinders as the surface loads are applied, it has been argued that the assumption of elastic conditions may not be too broad. In fact, the experimental results support this view.

The experiments were mainly model tests in laboratory soils, and there is little evidence to indicate how closely the results can be applied to full size structures under field conditions. However, it is believed that a design based on the equations in the text, used with the proper safety factors, would be reasonably conservative. At the scales used in the tests the deadweight of the soil had very little influence on the measurements, but in applying the results to full-scale designs, the deadweight of the soil backfill should be taken into account as an additional overpressure.

It is clear that thin-walled cylindrical tubes, carefully bedded in soils so that initial deviations from the geometry are low, are capable of surviving without collapse under high surface loads. There seems little to be gained in burying the tubes more than one diameter deep (surface to crown). The effects of dynamic surface pressure are being actively pursued, and test results have been presented by the author and others. As yet, however, the analysis is not sufficiently developed, and future work in this field will be aimed at establishing suitably concise design procedures.

REFERENCES

1. TIMOSHENKO, S. P. and GERE, J. M., *Theory of Elastic Stability*, McGraw-Hill, New York, 1961, p. 292.
2. CHENEY, J. A., *Proc. ASCE*, **89**, EM5, October 1963.
3. MEYERHOF, G. G. and BAIKIE, L. D., Strength of steel culvert sheets bearing against compacted sand backfill, *Highway Research Record, No. 30*, 1963.
4. LUSCHER, U. and HÖEG, K., The interaction between a structural tube and the surrounding soil, *Air Force Weapons Lab.*, *Report RTD TDR-63-3109*, Kirtland, New Mexico, 1964.
5. FORRESTAL, M. J. and HERRMANN, G., *Int. J. Solids. Struct.*, **1**, 297, 1965.
6. DUNS, C. S., The elastic critical load of a cylindrical shell embedded in an elastic medium, *Report CE/10/66*, University of Southampton, 1966.
7. HABIB, P. and PHONG, L. M., *Annales de l'Institut Technique de Batiment et des Travaux Publics No. 218*, February 1966.
8. SONNTAG, G., *Die Stabilität dunnwardoger Rohve in Noträsionslosen Kantinunon, Rockmechanics und Engineering Geology*, Vol. IV/3, Springer Verlag, Berlin, 1966.
9. CHELEPATI, C. V., Critical pressures for radically-supported cylinders, *Tech. Note N-773, US Naval Civil Eng. Lab.*, California, January 1966.
10. COMPSTON, D. G., CRAY, P., SCHOFIELD, A. N. and SHANN, C. D., Design and construction of buried thin-wall pipes, *CIRIA Report 78*, London, 1978.
11. ALLGOOD, J. R. and CIANI, J. B., The influence of soil modulus on the behaviour of cylinders in sand, *Highway Research Record, No. 249*, Washington, 1968, p. 1.
12. BULSON, P. S., Thin-walled tubular structures under soil cover and surface pressure, *Report 75.15/18, Military Vehicles and Engineering Establishment*, Christchurch, UK, 1972.
13. CHENEY, J. A., Buckling of the thin-walled cylindrical shell in soil, *Cambridge University Engineering Dept., Report CVED/C—SOILS TR 26*, 1976.
14. WATKINS, R. K., *Proc. Symp. on Soil Structure Interaction*, University of Arizona, June 1964.
15. WHITE, H. L. and LAYER, J. P., *Proc. Highway Research Board*, **39**, 389, 1960.

16. LUSCHER, U., *Proc. ASCE.*, *SM6*, 211, November 1966.
17. BULSON, P. S., Stability of buried tubes under static and dynamic overpressure; part 1: circular tubes in compacted sand, *Research Report RES 47.5/7, Military Vehicles and Engineering Establishment*, Christchurch, UK, 1966.
18. DORRIS, A. F., Response of horizontally orientated buried cylinders to static and dynamic loading, *US Army Engineer Waterways Experiment Station Tech. Report No. 1-682*, July 1965.
19. BULSON, P. S., *Proc. Symp. of Soil–Structure Interaction*, University of Arizona, USA, June 1964, p. 211.
20. BULSON, P. S., Static and dynamic overpressure on brittle circular tubes in clay, *Research Report 47.5/15, Military Vehicles and Engineering Establishment*, Christchurch, UK, June 1969.
21. ALLISON, C. J., An experimental study of the strength of circular tubes buried in a confined medium, *Research Report 47.5/8, Military Vehicles and Engineering Establishment*, Christchurch, UK, February 1967.
22. ENGLISH, R. J. and SCHOFIELD, A. N., *Geotechnique*, **24**(1), 1974.
23. ENGLISH, R. J., *PhD Thesis*, Manchester University, 1973.
24. VALSANGKAR, A. J. and BRITTO, A. M., Centrifuge tests of flexible circular pipes subjected to surface loading, *TRRL Supplementary Report 530*, 1979.
25. VALSANGKAR, A. J. and BRITTO, A. M., The validity of ring compression theory in the design of flexible buried pipes, *TRRL Report SR 440*, 1978.
26. HOWARD, A. K., Laboratory load tests on buried flexible pipe, *Progress Report No. 1, Soils Engineering Branch, Bureau of Reclamation, US Dept. of the Interior Report EM-763*, June 1968.
27. PETTIBONE, M. C. and HOWARD, A. K., Laboratory investigation of soil pressures on concrete pipe, *Progress Report No. 2, Soils Engineering Branch, Bureau of Reclamation, US Dept. of the Interior, Report EM-718*, March 1966.
28. MCCAVOUR, T. C., Composite design for soil-steel structures, *Corrugated Steel Pipe Institute Technical Bulletin No. 205*, 1968.
29. MEYERHOF, G. G., *Canadian Good Roads Association, Annual Convention*, September 1966.
30. ALLGOOD, J. R., *Proc. Symp. of Soil–Structure Interaction*, University of Arizona, 1964, p. 189.
31. DONNELLAN, B. A., *Proc. Symp. of Soil–Structure Interaction*, University of Arizona, 1964, p. 449.
32. WHITE, H. L., A rational approach to soil pressures surrounding flexible metal structures, *ARMCO publication*, 1957.
33. GILL, H. L., Active arching of sand during dynamic loading; results of an experimental program and development of analytical procedure, *US Naval Civ. Eng. Lab., Tech. Rep. R541*, Port Hueneme, California, September 1967.

Chapter 3

DYNAMIC BEHAVIOUR OF A THIN-WALLED CIRCULAR CYLINDRICAL SHELL

N. Yamaki

Institute of High Speed Mechanics, Tohoku University,
Sendai, Japan

SUMMARY

In this chapter, three fundamental problems are discussed concerning the dynamic behaviour of a thin-walled circular cylindrical shell of elastic material. These are linear free vibrations, non-linear vibrations with large amplitudes, and parametric instability under axial excitation. For the first and the third topics, emphasis is placed on revealing the general characteristics of the relevant solutions while for the second topic only the methods of solution are presented for the problems commonly covered in experimental work. Results of the corresponding experimental studies are also presented, which seem to supplement the foregoing analyses and provide new and reliable data facilitating further theoretical researches.

1. INTRODUCTION

It is of great technical importance to clarify the dynamic characteristics of the thin-walled circular cylindrical shell, as it constitutes a basic element in modern light-weight structures. In this chapter, we shall study the three fundamental topics related to this subject; these are, linear free vibrations, non-linear vibrations with large amplitudes and dynamic stability under periodic axial loading.

In the first part of this chapter, we shall consider the free vibration problem with small amplitudes. Although this problem has been solved

81

accurately for various boundary conditions, actual calculations are confined to some special cases owing to the computational complexity. Hence, the emphasis is placed on illustrating the general aspect of the free vibration characteristics for a wide range of shell geometries, under example boundary conditions where both edges are completely clamped. In the second part of the chapter, we shall study the non-linear free vibration problem with large amplitudes. This is one of the long debated problems in applied mechanics for which no satisfactory agreement has yet been achieved between theory and experiment. As a first step to overcome the difficulty, the problem is properly formulated and then appropriate methods of solution are presented: one based on a direct non-linear analysis, the other utilising a perturbation procedure. Actual calculations are left for future studies.

In the third section of the chapter we shall consider the parametric excitation problem, that is, the dynamic stability under periodic axisymmetric loading. Considering that relatively few researches have been conducted on this problem, we shall outline the method of solution and then illustrate the general features of the instability regions, taking the typical case when the shell is subjected to periodic compressive forces applied along the edges.

In the final part, we shall illustrate the results of experimental studies recently conducted to supplement the preceding theoretical analyses. A polyester test cylinder with radius 100 mm, length 226 mm and thickness 0·25 mm was used and the shell responses were observed with a non-contacting displacement transducer called a 'Fotonic Sensor'. For the free vibration, the shell was laterally excited at some point along the surface while for the parametric excitation, the shell was almost filled with water and was subjected to a periodic compressive force applied at the centre of the upper cover plate of the cylinder. Test results here obtained for the linear free vibration are found to be in excellent agreement with theoretical predictions while those for the non-linear vibration and parametric excitation seem to provide new and reliable data for further theoretical studies in the future.

2. LINEAR FREE VIBRATION

One of the most fundamental problems in the dynamics of structures is that of clarifying the free vibration characteristics of circular cylindrical shells. Hence, numerous researches have been conducted on this subject since the linear shell theory was established by Flügge[1] in 1932. In the earlier stages the problem was solved approximately under special boundary conditions and accurate solutions under various boundary conditions were first obtained by Forsberg[2] only in 1964, with the advent of a high-speed digital computer. The problem was further pursued by Warburton,[3] Forsberg,[4] Vronay and Smith,[5] Warburton and Higgs[6] and Goldman.[7] Based on the linear theory derived by Sanders,[8] the problem was re-examined by Dym[9] under the classical simply supported conditions, showing the results to be coincident with those predicted by the Flügge theory. Recently, approximate methods of solution for this problem were presented by Sharma[10,11] and Soedel.[12]

Through the studies mentioned above, the present problem would appear to have been fully explored. However, owing to inherent computational difficulties as well as the complexity of the subject, actual calculations were carried out only for some particular combinations of the shell geometric parameters, from which it is difficult to grasp the general features of the problem. Hence, it is intended here to illustrate the whole aspect of the free vibration characteristics of circular cylindrical shells, for the typical case where both edges are completely clamped. For that purpose, we restrict the problem to the lower order natural vibrations of thin shells where the vibration is predominantly flexural. In this case, we can apply, instead of the Flügge equations, the dynamic version of the Donnell equations[13] neglecting the effect of in-plane inertia forces, which facilitates the compact representation of the results as will be seen in the following.[14]

Now we shall outline the method of solution for the linear free vibration of a shell with radius R, length L and thickness h. Taking the coordinate system as shown in Fig. 1 and denoting by U, V, W the displacement components, the Donnell basic equations are given by

$$\left. \begin{array}{l} U_{,xx} + \dfrac{1-v}{2} U_{,yy} + \dfrac{1+v}{2} V_{,xy} - \dfrac{v}{R} W_{,x} = 0 \\[2ex] \dfrac{1+v}{2} U_{,xy} + \dfrac{1-v}{2} V_{,xx} + V_{,yy} - \dfrac{1}{R} W_{,y} = 0 \\[2ex] -\dfrac{v}{R} U_{,x} - \dfrac{1}{R} V_{,y} + \dfrac{1}{R^2} W + (1-v^2) \dfrac{D}{Eh} \nabla^4 W + (1-v^2) \dfrac{\rho}{E} W_{,tt} = 0 \end{array} \right\} \quad (1)$$

where

$$D = \frac{Eh^3}{12(1-v^2)} \qquad \nabla^2 = \frac{\partial^2}{\partial x^2} + \frac{\partial^2}{\partial y^2} \qquad (2)$$

In these equations, D is the flexural rigidity, E, v, ρ are Young's modulus, Poisson's ratio and the density of the shell, respectively, t is the time, while

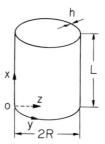

FIG. 1.   Shell geometry and the coordinate system.

subscripts following a comma stand for differentiation. For the shell with both edges completely clamped, the boundary conditions become

$$x = 0 \qquad L: W = W_{,x} = U = V = 0 \qquad (3)$$

To solve the problem, we assume the solution in the form

$$\left. \begin{aligned} U &= (Lh/R)A \exp{(rx/L)} \cos{(Ny/R)} \cos{\Omega t} \\ V &= (Lh/R)B \exp{(rx/L)} \sin{(Ny/R)} \cos{\Omega t} \\ W &= hC \exp{(rx/L)} \cos{(Ny/R)} \cos{\Omega t} \end{aligned} \right\} \qquad (4)$$

where A, B, C and r are complex constants, N is the circumferential wave number while Ω is the natural frequency. Substituting these expressions into eqns (1), we obtain a set of three homogeneous linear equations in A, B and C. In order that these equations have non-trivial solutions, the determinant of the coefficients should vanish, from which we obtain the eighth order algebraic equation in r as the characteristic equation. In the meantime, we obtain the expressions for $\phi = A/C$ and $\psi = B/C$ from the foregoing set of equations. Hence, denoting by r_j $(j = 1-8)$ the distinct roots of the characteristic equation, the general solution can be expressed as

$$\left. \begin{aligned} U &= (Lh/R) \sum_{j=1}^{8} C_j \phi_j \exp{(r_j x/L)} \cos{(Ny/R)} \cos{\Omega t} \\ V &= (Lh/R) \sum_{j=1}^{8} C_j \psi_j \exp{(r_j x/L)} \sin{(Ny/R)} \cos{\Omega t} \\ W &= h \sum_{j=1}^{8} C_j \exp{(r_j x/L)} \cos{(Ny/R)} \cos{\Omega t} \end{aligned} \right\} \qquad (5)$$

where C_j values ($j = 1–8$) are the unknown constants. With these expressions and the boundary conditions, eqns (3), we have a set of eight homogeneous linear equations in C_j. For non-trivial solutions, the determinant of the coefficients, Δ, should vanish, which gives the frequency equation of the problem. Here, it is to be noted that Δ in this case becomes a function of Poisson's ratio v and the three non-dimensional parameters Z, β and ω, relating to the shell geometry, wave number and the natural frequency, respectively. Thus, we have

$$\Delta(v, Z, \beta, \omega) = 0 \tag{6}$$

where

$$Z = \sqrt{1 - v^2} \, \frac{L^2}{Rh} \qquad \beta = \frac{L}{\pi R} \, N \qquad \omega = \frac{L^2}{\pi^2} \sqrt{\frac{\rho h}{D}} \, \Omega \tag{7}$$

Hence, when the values of v and Z are given, we can determine, for any prescribed value of β, values of ω successively in the order of smallness, which give the natural frequencies ω_m, where $m = 1, 2, 3 \ldots$. The corresponding mode of vibration can be determined by substituting the values of β and ω_m into the original set of equations in C_j. It will be seen that the mth mode of vibration with frequency ω_m has m half-waves in the axial direction and is symmetric or antisymmetric with respect to the central section of the shell, according to whether m is odd or even. The natural frequency ω_m obviously depends on the wave number β and the minimum natural frequency and the corresponding wave number for the specified shell will be denoted by $\omega_{m(0)}$ and $\beta_{m(0)}$, respectively.

Based on the foregoing analysis, values of $\omega_{m(0)}$ and $\beta_{m(0)}$, where $m = 1–6$, are determined for a wide range of the shell geometry, Z, with the result as shown in Fig. 2. Poisson's ratio v has been assumed to be 0·3. In the figure, m and N are the order of the axial mode and the circumferential wave number, respectively, as stated before. For relatively long shells, we have the approximate expression for $\omega_{m(0)}$ where $m = 1–6$,

$$\omega_{m(0)} = K_m (m + 0·5) \sqrt{Z} \qquad \text{for } Z \geq 500 \tag{8}$$

where $0·76 \leq K_m \leq 0·82$. To illustrate the variations of the natural frequency, ω_m, with the wave number, β_m, the relation between $\omega_m / \omega_{m(0)}$ and $\beta_m / \beta_{m(0)}$ is plotted in Fig. 3, where the effect of Z on the first mode of vibration is depicted in (a) while the corresponding effects on the first six modes of vibration are illustrated in (b) and (c), for the cases when $Z = 10^2$ and 10^4, respectively.

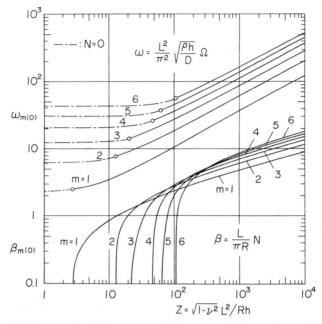

FIG. 2. Minimum natural frequency and the corresponding wave number for the completely clamped cylindrical shell: $v = 0.3$.

When the shell dimensions and material properties are given, we can predict, from Figs 2 and 3, the natural frequencies and the corresponding wave numbers for the first six modes of vibration, representing the global aspect of the free vibration characteristics of the shell. For instance, the test cylinder used in the following experimental study has the dimensions $R = 100$ mm, $L = 226.0$ mm, $h = 0.247$ mm with $v = 0.3$, leading to $Z = 1973$. Values of $\omega_{m(0)}$ and $\beta_{m(0)}$ for the shell with $Z = 2000$, corresponding to $m = 1$, 2 or 3 are 53.40, 88.75 or 124.6 and 5.96, 7.52 or 8.74, respectively.[14] Considering that for large values of Z, $\omega_{m(0)}$ and $\beta_{m(0)}$ vary in proportion to $Z^{1/2}$ and $Z^{1/4}$, respectively, we will find the values of $\omega_{m(0)}$ and $\beta_{m(0)}$ for $Z = 1973$ as 53.03, 88.14, 123.7 and 5.94, 7.49, 8.71, respectively. Variations of the natural frequencies with the wave number can be estimated from Fig. 3, while the corresponding physical quantities will be obtained by specifying the material properties E, v and ρ. Comparison with experiment will be considered later in the chapter.

Finally, we shall examine the validity of present solutions based on the Donnell equations neglecting the effect of in-plane inertia forces. The

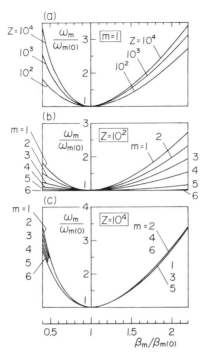

FIG. 3.   Variations of the natural frequency ω_m with the wave number β_m for the completely clamped cylindrical shell: $\nu = 0\cdot3$.

accuracy of the Flügge and Donnell equations has been studied by Armenàkas[15] through a numerical comparison of their results with those obtained by elasticity theory. It is found that the Flügge equations are sufficiently accurate as long as the shell is thin and the vibration is predominantly flexural, while the Donnell equations may lead to a significant error when the shell is long and the circumferential wave number is small, i.e. when $L/R \gg 1$ and $N \leq 4$. Further, the effect of in-plane inertia forces is found negligible when the shell is thin and N is larger than 3. Judging from the foregoing together with the relations between Z and $\beta_{m(0)}$ here obtained, the present results are estimated to be valid for thin shells with both L/h and R/h greater than 50, as long as the values of Z are less than 10^3, 10^4 and 10^5, respectively, corresponding to the cases when R/h values are 100, 300 and 1000. It should be noted that in reference 14, similar calculations have been carried out for eight different boundary conditions, clarifying the effects of boundary conditions on the present problem.

3. NON-LINEAR VIBRATIONS

When the vibration amplitude becomes comparable to or larger than the shell thickness, the non-linear interaction between the bending and stretching of the shell becomes significant leading to a variety of complex responses which cannot be analysed within the linear theory of shells. The problem then should be treated with the non-linear theory of shells, taking the effects of large amplitudes into consideration. With the recent trend to use increasingly thin shell structures in various industrial fields, it will be of great technical importance to clarify the non-linear response of thin-walled circular cylindrical shells. Hence, numerous researches have been conducted on this subject since the pioneering work of Reissner[16] in 1955. A detailed historical review was presented by Evensen[17] concerning research carried out before 1972. Most of the analytical studies were based on the Donnell non-linear theory of circular cylindrical shells,[13] while a few experimental studies were conducted using extremely thin cylindrical shells. It was concluded that certain fundamental questions still persisted for which precise experimental data were much needed.

We shall now review the relevant studies since 1972. Atluri[18] proposed a method of solution for this problem in which the Galerkin procedure is first used with the Donnell basic equations and then the method of multiple timescales is applied to the resulting set of coupled non-linear differential equations. With this method, a non-linear hardening effect was found for the large amplitude free vibration of simply supported cylinders with the edges axially constrained. A perturbation procedure analogous to those used for the initial post-buckling problem by Koiter[19] and Budiansky[20] was derived by Rehfield,[21,22] which was applied to non-linear vibrations of beams and plates. Based on the non-linear shell theory of Sanders,[23] general modal equations for the non-linear vibration of shells were obtained by Radwan and Genin,[24] which were then specialised for the simply supported cylindrical shell.[25] Applying a perturbation procedure to the Donnell non-linear equations, Chen and Babcock[26] obtained approximate solutions for the present problem under the classical simply supported conditions, indicating that the non-linearity of the frequency–response relation was either softening or hardening depending on the mode of vibration. The corresponding experimental studies were conducted by acoustically exciting a thin test cylinder, the results of which were found to be in qualitative agreement with the theoretical ones. Travelling wave phenomena and some non-stationary motions were also observed at the resonant responses. Recently, a general perturbation procedure was proposed by Harari[26] to analyse free vibrations of pre-stressed plates and shells. The analysis is based on the non-linear equations of motion by Sanders[23] including the effect of in-plane inertia forces and no assumption is made *a priori* of the form of the time and space mode. The applicability was demonstrated by analysing the non-linear vibration of the infinite strip under various boundary conditions. Further, applying the finite element method, the present problem has been studied by Raju and Rao[27] and by Ueda,[28] as the special cases of the non-linear vibrations of shells of revolution and conical shells, respectively.

Even with all the foregoing research, it is to be noted that the present problem has not been fully explored, leaving the 'fundamental questions' posed by Evensen[17] still unsolved. To clarify the whole aspect of the present problem, it seems obvious that more accurate analyses as well as more precise experiments should be performed systematically, until satisfactory agreement is confirmed between theory and experiment. As a first step in tackling the difficulty, we shall present here a proper formulation of the present problem and then outline two promising methods of solution; that is, one based on the direct non-linear analysis and the other utilising a perturbation procedure. Actual calculations are left for future study. Results of a preliminary experimental study will be stated later.

Now, assume that a circular cylindrical shell with radius R, length L and thickness h is executing free vibrations with large amplitude (Fig. 1). Confining the problem to predominantly flexural vibrations of sufficiently thin cylinders, effects of the transverse shear deformation, rotatory inertia as well as in-plane inertia forces can be neglected and the governing equations may be given by the well known Donnell non-linear theory as

$$\nabla^4 F + Eh(R^{-1}W_{,xx} + W_{,xx}W_{,yy} - W_{,xy}^2) = 0 \quad (9)$$

$$D\nabla^4 W - R^{-1}F_{,xx} - F_{,xx}W_{,yy} + 2F_{,xy}W_{,xy} - F_{,yy}W_{,xx} + \rho h W_{,tt} = 0 \quad (10)$$

$$\nabla^2 = \partial^2/\partial x^2 + \partial^2/\partial y^2 \qquad D = Eh^3/12(1-v^2) \quad (11)$$

where W and F are the normal displacement (positive inward) and the stress function, respectively. Other notations are the same as defined in the previous section. The stress resultants are expressed by the stress function as

$$N_x = F_{,yy} \qquad N_y = F_{,xx} \qquad N_{xy} = -F_{,xy} \quad (12)$$

while the in-plane displacements U and V are related to F and W as

$$\left. \begin{array}{c} Eh(U_{,x} + \tfrac{1}{2}W_{,x}^2) = F_{,yy} - vF_{,xx} \\ Eh(V_{,y} - R^{-1}W + \tfrac{1}{2}W_{,y}^2) = F_{,xx} - vF_{,yy} \\ Eh(U_{,y} + V_{,x} + W_{,x}W_{,y}) = -2(1+v)F_{,xy} \end{array} \right\} \quad (13)$$

In the experiment, it is usual that both edges are connected to stiff end plates, the inertia forces of which are negligibly small. Hence, we have the boundary conditions at $x = 0$, L as

$$\left.\begin{array}{c} W = W_{,x} = 0 \\[2mm] U_{,y} = V_{,y} = \displaystyle\int_0^{2\pi R} N_{,x}\,dy = \int_0^{2\pi R} N_{,xy}\,dy = 0 \end{array}\right\} \quad (14(a))$$

or

$$F_{,xx} - vF_{,yy} = F_{,xxx} + (2+v)F_{,xyy} = [F_{,x}]_{y=0}^{2\pi R} = [F_{,y}]_{y=0}^{2\pi R} = 0 \quad (14(b))$$

Here, assuming that the vibration takes place in N circumferential waves with a natural frequency Ω, we introduce the following notations for convenience:

$$\left.\begin{array}{l} \xi = \pi x/L \quad \eta = \pi y/l \quad l = \pi R/N \quad \beta = L/l = (L/\pi R)N \\[2mm] c = 1/12(1-v^2) \quad (u,v) = (L/\pi h^2)(U,V) \quad w = W/h \quad f = F/Eh^3 \\[2mm] (n_x, n_y, n_{xy}) = (L^2/\pi^2 Eh^3)(N_x, N_y, N_{xy}) \quad \alpha = L^2/\pi^2 Rh \\[2mm] Z = \sqrt{1-v^2}\, L^2/Rh \quad \tau = \Omega t \quad \Gamma = (\rho h^2 L^4/\pi^4 Eh^4)\Omega^2 = c\omega^2 \end{array}\right\} \quad (15)$$

In the foregoing, α and Z are geometrical parameters of the shell, β is a wave number parameter while ω and Γ are non-dimensional parameters relating to the natural frequency. With this notation, the preceding equations are rewritten as follows:

$$\bar{\nabla}^4 f + \alpha w_{,\xi\xi} + \tfrac{1}{2}\beta^2(w;w) = 0 \tag{9'}$$

$$L(w) \equiv c\bar{\nabla}^4 w - \alpha f_{,\xi\xi} - \beta^2(f;w) + \Gamma w_{,\tau\tau} = 0 \tag{10'}$$

$$\bar{\nabla}^2 = (\partial^2/\partial\xi^2) + \beta^2(\partial^2/\partial\eta^2) \tag{11'}$$

$$n_x = \beta^2 f_{,\eta\eta} \quad n_y = f_{,\xi\xi} \quad n_{xy} = -\beta f_{,\xi\eta} \tag{12'}$$

$$\left.\begin{array}{l} u_{,\xi} + \tfrac{1}{2}w_{,\xi}^2 = \beta^2 f_{,\eta\eta} - vf_{,\xi\xi} \\[2mm] \beta v_{,\eta} - \alpha w + \tfrac{1}{2}\beta^2 w_{,\eta}^2 = f_{,\xi\xi} - v\beta^2 f_{,\eta\eta} \\[2mm] \beta u_{,\eta} + v_{,\xi} + \beta w_{,\xi} w_{,\eta} = -2(1+v)\beta f_{,\xi\eta} \end{array}\right\} \tag{13'}$$

$$\xi = 0 \quad \pi: M_i(w,f) = 0 \quad i = 1\text{–}6 \tag{14'}$$

In these equations, we have introduced the following notation:

$$
\left.
\begin{aligned}
&(f;w) = f_{,\xi\xi}w_{,\eta\eta} - 2f_{,\xi\eta}w_{,\xi\eta} + f_{,\eta\eta}w_{,\xi\xi} \\
&M_1(w,f) = w \qquad M_2(w,f) = w_{,\xi} \qquad M_3(w,f) = f_{,\xi\xi} - v\beta^2 f_{,\eta\eta} \\
&M_4(w,f) = f_{,\xi\xi\xi} + (2+v)\beta^2 f_{,\xi\eta\eta} \qquad M_5(w,f) = [f_{,\xi}]_{\eta=0}^{2\pi} \\
&M_6(w,f) = [f_{,\eta}]_{\eta=0}^{2\pi}
\end{aligned}
\right\}
\tag{16}
$$

Further, both w and f should be periodic functions in τ with period 2π. Hence

$$
w(\xi,\eta,\tau+2\pi) - w(\xi,\eta,\tau) = 0 \qquad f(\xi,\eta,\tau+2\pi) - f(\xi,\eta,\tau) = 0
$$

which will be symbolised as

$$
P_\tau(w,f:2\pi) = 0 \tag{17}
$$

The problem consists in determining the steady-state solutions for w, f and Γ from these equations, when the values for Poisson's ratio v, shell geometry Z (or α) and wave number β (or N) are prescribed.

We shall now proceed to the method of solution. The first one is based on direct non-linear analysis. In this case, we assume the solution w as

$$
\left.
\begin{aligned}
&w = \sum_{m=1}\sum_{n=0} a_{mn}(\tau)w_{mn}(\xi,\eta) \\
&w_{mn} = [\cos(m-1)\xi - \cos(m+1)\xi]\cos n\eta \\
&\qquad m = 1,2,3\ldots, \ n = 0,1,2\ldots
\end{aligned}
\right\}
\tag{18}
$$

where $a_{mn}(\tau)$ are the unknown time functions and where $w_{mn}(\xi,\eta)$ are the spatial functions satisfying the first two of the boundary conditions, eqns (14'). It is to be noted that w_{mn} values with m odd and n even, correspond to the symmetric and antisymmetric axial modes, respectively, while those with $n = 0, 1, 2\ldots$ represent the axisymmetric, fundamental, second and higher harmonic modes in the circumferential direction, respectively. Substituting eqns (18) into eqn (9') and integrating, we obtain the general solution f as $f = f_p + f_c$, where f_p is a particular integral containing both linear and quadratic terms in $a_{mn}(\tau)$ and where f_c is the complementary function which should be determined so that the remaining four of the boundary conditions, eqns (14'), are satisfied. With the functions w and f thus determined, we apply the Galerkin method to the remaining basic eqn (10'), which leads to the following conditions for each spatial function w_{rs}:

$$\int_0^\pi \int_0^{2\pi} L(w) w_{rs}(\xi, \eta)\, d\xi\, d\eta = 0 \qquad r = 1, 2, 3 \ldots \qquad s = 0, 1, 2 \ldots \qquad (19)$$

These equations represent a set of coupled non-linear differential equations in $a_{mn}(\tau)$. Rewriting $a_{mn}(\tau)$ with $a_i(\tau)$, $i = 1, 2, 3 \ldots$, for convenience, we have the equations in the form

$$\Gamma A_{ij} a_{j,\tau\tau} + B_{ij} a_j + C_{ijk} a_j a_k + D_{ijkl} a_j a_k a_l = 0 \qquad i, j, k, l = 1, 2, 3 \ldots \qquad (20)$$

where the usual summation convention has been used and where A_{ij}, B_{ij}, C_{ijk} and D_{ijkl} are the coefficients depending on the parameters v, Z and β. The linear free vibration problem can be solved easily with these equations, omitting the non-linear terms in $a_i(\tau)$. To solve the non-linear vibration problem, we assume the solution $a_i(\tau)$ as

$$a_i(\tau) = \sum_j b_{ij} \cos j\tau \qquad j = 0, 1, 2 \ldots \qquad (21)$$

where b_{ij} are unknown constants. It will be noted that b_{ij} with $j = 0, 1, 2 \ldots$ represent the time-independent, fundamental, second and higher harmonic components of $a_i(\tau)$, respectively. Substituting these expressions into eqn (20) and applying the harmonic balance method, we will obtain a set of simultaneous cubic equations in b_{ij}. Solving these for b_{ij} in connection with Γ, the corresponding expressions for w and f become known and the problem will be solved. The accuracy of the solution thus obtained depends on the choice and number of unknown functions, $a_{mn}(\tau)$, as well as the unknown constants, b_{ij}, taken into consideration.

A similar procedure consisting of successive application of the Galerkin and harmonic balance methods has been applied to analyse the non-linear forced vibration problems of the clamped beam[29] and the clamped circular plate,[30] the results of which were found to be in reasonable agreement with the corresponding experimental ones. A static counterpart of the present problem, i.e. the postbuckling problem of the clamped cylindrical shell, has been solved with the Galerkin method under various loading conditions,[31,32] the results of which were confirmed to be in good agreement with the corresponding experimental ones.[33,34]

As the second method of solution, we now outline a perturbation procedure applied to the foregoing Donnell basic equations. First, we assume the solution to be expanded as

$$
\left.\begin{array}{l}
w = w(\xi, \eta, \tau, \varepsilon) = \varepsilon w_1 + \varepsilon^2 w_2 + \varepsilon^3 w_3 + \cdots \\
f = f(\xi, \eta, \tau, \varepsilon) = \varepsilon f_1 + \varepsilon^2 f_2 + \varepsilon^3 f_3 + \cdots \\
\Gamma = \Gamma(\varepsilon) = \Gamma_0 + \varepsilon \Gamma_1 + \varepsilon^2 \Gamma_2 + \cdots
\end{array}\right\} \tag{22}
$$

where ε is a perturbation parameter. Imposing the condition

$$
\frac{1}{L_{\xi\eta\tau}} \int_0^\pi \int_0^{2\pi} \int_0^{2\pi} w w_1 \, d\xi \, d\eta \, d\tau = \varepsilon
$$

we have the orthogonality condition

$$
\frac{1}{L_{\xi\eta\tau}} \int_0^\pi \int_0^{2\pi} \int_0^{2\pi} w_1 w_n \, d\xi \, d\eta \, d\tau = \delta_{1n} \qquad n = 1, 2, 3 \ldots \tag{23}
$$

where $L_{\xi\eta\tau}$ is an arbitrary constant related to the domain of integration and δ_{1n} is the Kronecker delta.

Substituting eqns (22) into eqns (9′), (10′), (14′) and (17) and retaining only the terms with ε, we obtain

$$
L_1(w_1, f_1) \equiv \bar{\nabla}^4 f_1 + \alpha w_{1,\xi\xi} = 0 \tag{24(a)}
$$

$$
L_2(w_1, f_1) \equiv c\bar{\nabla}^4 w_1 - \alpha f_{1,\xi\xi} + \Gamma_0 w_{1,\tau\tau} = 0 \tag{24(b)}
$$

$$
\xi = 0 \qquad \pi : M_i(w_1, f_1) = 0 \qquad i = 1\text{–}6 \tag{24(c)}
$$

$$
P_\tau(w_1, f_1 : 2\pi) = 0 \tag{24(d)}
$$

To solve the eigenvalue problem posed in the foregoing, we assume the solution w_1 to be

$$
w_1 = \sum_m a^{(1)}_{m11} w_{m1}(\xi, \eta) \cos \tau
$$

$$
= \sum_m a^{(1)}_{m11} [\cos(m-1)\xi - \cos(m+1)\xi] \cos \eta \cos \tau
$$

$$
m = 1, 2, 3 \ldots \tag{25}
$$

where $a^{(1)}_{m11}$ are unknown constants and where $w_{mn}(\xi, \eta)$ have been defined in eqn (18). Substituting these into eqn (24(a)) and considering the conditions eqns (24(c)) and (24(d)), we can determine the stress function f_1

in terms of the unknown constants $a_{m11}^{(1)}$. With the expressions w_1 and f_1 thus obtained, we apply the Galerkin method to eqn (24(b)), which leads to a set of linear homogeneous equations in $a_{m11}^{(1)}$. The eigenvalue Γ_0 and the corresponding eigenfunctions w_1 and f_1 can be determined from these equations by normal procedures. The expressions for w_1 and f_1 are determined from the orthogonality condition, eqn (23).

Here, we note that the present eigenvalue problem is equivalent to the variational relation

$$\int_0^\pi \int_0^{2\pi} \int_0^{2\pi} [L_1(w_1,f_1)\delta f - L_2(w_1,f_1)\delta w]\,d\xi\,d\eta\,d\tau = 0$$

where δw and δf are arbitrary variations for w_1 and f_1, respectively. Putting $\delta w = w_n$ and $\delta f = f_n$ in this expression and transforming, we obtain the so-called solvability conditions

$$\int_0^\pi \int_0^{2\pi} \int_0^{2\pi} [L_1(w_n,f_n)f_1 - L_2(w_n,f_n)w_1]\,d\xi\,d\eta\,d\tau = 0 \qquad n = 2, 3, 4 \ldots$$

$$(26)$$

which also serve to determine the parameters $\Gamma_n, n = 1, 2, 3 \ldots$, successively.

Next, retaining the terms with ε^2 in the aforementioned process, we have a set of equations as

$$L_1(w_2,f_2) = -\tfrac{1}{2}\beta^2(w_1; w_1) \qquad (27(a))$$

$$L_2(w_2,f_2) = \beta^2(f_1; w_1) - \Gamma_1 w_{1,\tau\tau} \qquad (27(b))$$

$$\xi = 0, \pi: M_i(w_2,f_2) = 0 \qquad i = 1\text{–}6 \qquad (27(c))$$

$$P_\tau(w_2,f_2: 2\pi) = 0 \qquad (27(d))$$

Application of the orthogonality condition, eqn (26), to the first two of these equations will lead to $\Gamma_1 = 0$. A pair of particular solutions of these equations, w_2^p and f_2^p, can be obtained through a similar procedure, by first assuming w_2^p as

$$w_2^p = \sum_m [(a_{m00}^{(2)} + a_{m02}^{(2)} \cos 2\tau)w_{m0} + (a_{m20}^{(2)} + a_{m22}^{(2)} \cos 2\tau)w_{m2}]$$

$$m = 1, 2, 3 \ldots \qquad (28)$$

where $a_{mnk}^{(2)}$ are unknown constants. The general solution will be given by

$$w_2 = w_2^p + c_2 w_1 \qquad f_2 = f_2^p + c_2 f_1 \qquad (29)$$

in which the unknown constant c_2 will be found to be zero from the orthogonality condition, eqn (23).

In the same manner, we obtain a set of equations with ε^3 as

$$L_1(w_3 \, f_3) = -\beta^2(w_1; w_2) \tag{30(a)}$$

$$L_2(w_3, f_3) = \beta^2(f_1; w_2) + \beta^2(f_2; w_1) - \Gamma_2 w_{1,\tau\tau} \tag{30(b)}$$

$$\xi = 0 \qquad \pi: M_i(w_3, f_3) = 0 \qquad i = 1\text{–}6 \tag{30(c)}$$

$$P_\tau(w_3, f_3: 2\pi) = 0 \tag{30(d)}$$

Application of the solvability condition, eqn (26), to these equations will yield the expression for Γ_2 as

$$\left.\begin{aligned}
\Gamma_2 &= \beta^2 B_2 / A_2 \\[4pt]
A_2 &= \int_0^\pi \int_0^{2\pi} \int_0^{2\pi} w_{1,\tau\tau} w_1 \, d\xi \, d\eta \, d\tau \\[4pt]
B_2 &= \int_0^\pi \int_0^{2\pi} \int_0^{2\pi} [(w_1; w_2)f_1 + (f_1; w_2)w_1 + (f_2; w_1)w_1] \, d\xi \, d\eta \, d\tau
\end{aligned}\right\} \tag{31}$$

After evaluating Γ_2, the particular solutions w_3^p and f_3^p can be obtained as before by assuming w_3^p in the form

$$w_3^p = \sum_m [(a_{m11}^{(3)} \cos \tau + a_{m13}^{(3)} \cos 3\tau)w_{m1} + (a_{m31}^{(3)} \cos \tau + a_{m33}^{(3)} \cos 3\tau)w_{m3}]$$

$$m = 1, 2, 3 \ldots \tag{32}$$

where $a_{mnk}^{(3)}$ are unknown constants. The general solution will be given by

$$w_3 = w_3^p + c_3 w_1 \qquad f_3 = f_3^p + c_3 f_1 \tag{33}$$

where the constant c_3 should be determined with the orthogonality condition, eqn (23).

The same procedure can be continued for the higher order solutions of w, f and Γ. Substituting these into eqns (22), we will obtain the solutions w, f and $\Gamma = cw^2 = (\rho h^2 L^4 / E h^4)\Omega^2$ in terms of ε, for the specific mode of natural vibration with wave number N. It is to be noted that the effect of amplitude on the natural frequency is of hardening or softening type according to whether Γ_2 in eqn (31) is positive or negative.

In contrast to the direct non-linear analysis, the perturbation method has the advantage of reducing the problem to a sequence of linear boundary value problems, which can be solved without difficulty. In general, however, the convergency of the solution is not good and the range of applicability is limited to some neighbourhood of the branching point. So far, the method has not actually been applied to the non-linear vibration problem of cylindrical shells. A similar perturbation method has been applied to the corresponding static problems, i.e. the post-buckling problems of the clamped cylinder under compression and under torsion.[35,36]

4. PARAMETRIC INSTABILITY

If a circular cylindrical shell is subjected to an axisymmetric periodic load, generally speaking, the shell will only undergo the axisymmetric forced vibration. Under certain dynamic load conditions, however, the axisymmetric response becomes unstable with a rapid build-up of a non-axisymmetric vibration leading to a violent flexural vibration of the shell. Dynamic instability of this kind is often called the parametric excitation problem, as the motion is governed by differential equations with time-varying coefficients. When this occurs, complex instability phenomena may arise and it is of considerable technical importance to clarify the general characteristics of the dynamic stability of circular cylindrical shells subjected to various axisymmetric excitation.

The basic reference on the parametric instability problem of elastic bodies is a book by Bolotin.[37] A variety of researches has been conducted concerning the present subject. In particular, the dynamic stability of simply supported cylinders under periodic axial and radial loads was treated by Bolotin,[37] Yao[38,39] and Wood and Koval,[40] while that of a vertical cylinder with one end clamped and the other end free, subjected to sinusoidal base motion, was studied by Vijayaraghavan and Evan-Iwanowski.[41] A general procedure for analysing the parametric instability of axisymmetric shells was proposed by Kalnins[42] and the effect of longitudinal resonance on dynamic stability was examined by Koval,[43] for simply supported cylindrical shells under axial excitation. Further, an excellent review article on the present topic together with the related ones was presented by Hsu.[44] Recently, the behaviour of shells subject to periodic shearing as well as compressive forces were studied by Yamaki and Nagai,[45,46] under four typical boundary conditions.

In most of the foregoing studies, the Donnell equations were used for the governing equations. In the earlier development, the unperturbed response was assumed to be of membrane stress state and only the principal and secondary instability regions were determined for the classical simply supported conditions. Recently, more accurate analyses have been conducted for various boundary conditions, considering the bending vibration in the unperturbed response and determining additional instability regions of the combination resonance type. Although the method of solution is well known, actual calculations were generally only carried out for illustration. Hence, we shall illustrate the general features of the instability regions in this section, taking the typical case when the completely clamped cylindrical shell is subjected to periodic compressive forces:[46]

Assume that a circular cylindrical shell with radius R, length L and thickness h is subjected to a uniformly distributed pulsating compressive load $q_0 + q_1 \cos \Omega t$ ($q_0 =$ intensity of static compressive load, $q_1 =$ amplitude of exciting compressive load, $\Omega =$ excitation frequency) along the edges $x = 0$ and L, Fig. 1. Restricting the problem to the low frequency responses of sufficiently thin shells, we shall employ the Donnell's non-linear shell theory as stated in the preceding section. Then the relevant equations for the axisymmetric unperturbed motion will be given by

$$N_{x0} = -(q_0 + q_1 \cos \Omega t) \qquad N_{y0} = v N_{x0} - EhR^{-1}W_0 \qquad (34)$$

$$DW_{0,xxxx} + q_0 W_{0,xx} + EhR^{-2}W_0 + \rho h W_{0,tt} + vR^{-1}(q_0 + q_1 \cos \Omega t) = 0 \qquad (35)$$

where $W_0(x,t)$ is the deflection and N_{x0} and N_{y0} are the components of stress resultants; other notations are the same as before. In eqn (35), the term $q_1 \cos \Omega t W_{0,xx}$ has been omitted to ease the analysis, which seems justifiable except for the case when the shell is extremely short.

Denoting the small incremental displacement components by U, V, W and the corresponding stress function by F, the governing equations for the perturbed motion will be obtained from the Donnell equations

$$\nabla^4 F = -Eh(R^{-1}W_{,xx} + W_{0,xx}W_{,yy}) \qquad (36)$$

$$D\nabla^4 W - R^{-1}F_{,xx} - (F_{,yy}W_{0,xx} + N_{x0}W_{,xx} + N_{y0}W_{,yy}) + \rho h W_{,tt} = 0 \qquad (37)$$

$$N_x = F_{,yy} \qquad N_y = F_{,xx} \qquad N_{xy} = -F_{,xy} \qquad (38)$$

$$\left.\begin{aligned} Eh(U_{,x} + W_{0,x}W_{,x}) &= F_{,yy} - vF_{,xx} \\ Eh(V_{,y} - R^{-1}W) &= F_{,xx} - vF_{,yy} \\ Eh(U_{,y} + V_{,x} + W_{0,x}W_{,y}) &= -2(1+v)F_{,xy} \end{aligned}\right\} \qquad (39)$$

Assuming that both edges are clamped to stiff end plates, the boundary conditions at $x = 0$ and L become

$$W_0 = W_{0,x} = 0 \qquad (40)$$

$$W = W_{,x} = U = V = 0 \qquad (41)$$

As stated in the foregoing, the unperturbed motion will become unstable under some load conditions. The problem consists of determining the instability regions around q_0, q_1 and Ω, when the shell dimensions as well as the material properties are specified.

We shall now outline the method of solution. To determine the unperturbed motion, we assume W_0 to be

$$W_0(x, t) = h \sum_m a_m(t)[\cos{(m - 1)}(\pi x/L) - \cos{(m + 1)}(\pi x/L)]$$

$$m = 1, 2, 3 \dots \qquad (42)$$

which satisfies the conditions of eqn (40). Applying the Galerkin method to eqn (35) and solving for the steady-state solution, we obtain the solution, $a_m(t)$, in the form

$$a_m(t) = a_{m0}(q_0) + q_1 a_{m1}(q_0)\cos \Omega t \qquad m = 1, 3, 5 \dots \qquad (42')$$

where a_{m0} and a_{m1} are constants depending on q_0. With these expressions, we have determined W_0 along with N_{x0} and N_{y0}.

Next, assuming that the perturbed motion has N circumferential wave number and considering the first two of the conditions of eqn (41), we put W into the following:

$$W(x, y, t) = h \sum_n b_n(t)[\cos{(n - 1)}(\pi x/L) - \cos{(n + 1)}(\pi x/L)\cos{(Ny/R)}]$$

$$n = 1, 2, 3 \dots \qquad (43)$$

where $b_n(t)$ are unknown time functions. With this expression, the stress function F can be determined in terms of $b_n(t)$ so as to satisfy the compatibility eqn (36) together with the remaining two conditions in eqn (41). With these expressions for W and F, we apply the Galerkin method to eqn (37), which finally yields a set of coupled Hill's equations in $b_n(t)$ in the form

$$\sum_n [A_{mn}b_{n,tt} + R_{mn}b_n - (P_{mn}q_1 \cos \Omega t + Q_{mn}q_1^2 \cos 2\Omega t)b_n] = 0$$

$$m, n = 1, 3, 5 \dots \text{ or } 2, 4, 6 \dots \qquad (44)$$

In these equations, odd and even integers for m and n correspond to the symmetric and antisymmetric axial modes of vibration, respectively, as can be seen in eqn (43).

Equation (44) involves those equations for the compressive buckling problem as well as those for the free vibration problem as the special cases. Hence, we can determine the buckling load q_{cr} and the corresponding wave number N_c as the lowest critical load of q_0. Meanwhile, for the prescribed wave number N, we can determine the sequence of natural frequencies $\bar{\Omega}_i$, $i = 1, 2, 3 \ldots$, together with the corresponding modes of vibration $\bar{W}_i$, taking the effect of static compressive load q_0 into consideration. Employing these data and transforming the generalised coordinates b_n to the normal coordinates c_i, we obtain, from eqn (44), the coupled Hill's equations in the standard form as

$$c_{i,\tau\tau} + \bar{\omega}_i^2 c_i - \bar{q}_1 \cos \omega\tau \sum_j \bar{P}_{ij} c_j - \bar{q}_1^2 \cos 2\omega\tau \sum_j \bar{Q}_{ij} c_j = 0$$

$$i, j = 1, 3, 5 \ldots, \text{ or } 2, 4, 6 \ldots \quad (45)$$

where we have introduced the following notations:

$$\left. \begin{aligned} \Omega_0 &= (\pi^2/L^2)\sqrt{D/\rho h} \qquad (\omega, \bar{\omega}_i) = (\Omega, \bar{\Omega}_i)/\Omega_0 \\ \tau &= \Omega_0 t \qquad (\bar{q}_0, \bar{q}_1) = (q_0, q_1)/q_{cr} \end{aligned} \right\} \quad (46)$$

Further, $\bar{P}_{ij}$ and $\bar{Q}_{ij}$ are constants depending on v, Z, β (see eqn (7)), ω and $\bar{q}_0$, which are symmetric with respect to the indices i and j.

The stability of the foregoing equations has been studied by Hsu[47] in detail. According to his first approximation analysis, eqn (45) has instability regions of combination resonance type when ω is in the neighbourhood of the sum of two natural frequencies $\bar{\omega}_i + \bar{\omega}_j$, the boundaries of which are given by

$$\left. \begin{aligned} \frac{\omega}{\omega_{ij}} &= 1 \pm \theta_{ij}\bar{q}_1 \\[2mm] \omega_{ij} &= \bar{\omega}_i + \bar{\omega}_j \\[2mm] \theta_{ij} &= \frac{|\bar{P}_{ij}|}{2(\bar{\omega}_i + \bar{\omega}_j)\sqrt{\bar{\omega}_i\bar{\omega}_j}} \qquad i, j = 1, 3, 5 \ldots, \text{ or } 2, 4, 6 \ldots \end{aligned} \right\} \quad (47)$$

In the foregoing, ω_{ij} and θ_{ij} are the central frequency and the relative width parameter of the instability region, respectively. It is to be noted that both

$\bar{\omega}_i$ and $\bar{\omega}_j$ correspond to the same wave number, N. Further, eqn (45) has the so-called principal instability regions at $\omega \simeq 2\bar{\omega}_i$, which are expressed as

$$\frac{\omega}{\omega_{ii}} = 1 \pm \theta_{ii}\bar{q}_1 \qquad \omega_{ii} = 2\bar{\omega}_i \qquad \theta_{ii} = \frac{|\bar{P}_{ii}|}{4\bar{\omega}_i^2} \qquad i = 1, 2, 3 \ldots \quad (48)$$

For a small excitation amplitude, $\bar{q}_1 = q_1/q_{cr}$, the foregoing instability regions will be usually most important. Through the second approximation analysis, the boundaries of the so-called secondary instability regions at $\omega \simeq \bar{\omega}_i$ are obtained as

$$1 - \theta_i^{(-)}\bar{q}_1^2 \le \frac{\omega}{\bar{\omega}_i} \le 1 + \theta_i^{(+)}\bar{q}_1^2$$

$$\left.\begin{array}{l}
\theta_i^{(+)} = \frac{1}{4\bar{\omega}_i^2}\left[\bar{Q}_{ii} - \sum_k \bar{P}_{ik}\bar{P}_{ki}/2(\bar{\omega}_k^2 - 4\bar{\omega}_i^2)\right] \\[4mm]
\theta_i^{(-)} = \frac{1}{4\bar{\omega}_i^2}\left\{\bar{Q}_{ii} + \sum_k \bar{P}_{ik}\bar{P}_{ki}\left[\frac{1}{\bar{\omega}_k^2} + \frac{1}{2(\bar{\omega}_k^2 - 4\bar{\omega}_i^2)}\right]\right\} \\[4mm]
i, k = 1, 3, 5\ldots, \text{ or } 2, 4, 6\ldots
\end{array}\right\} \quad (49)$$

Now we shall illustrate the results for the typical case when $Z = 100$, $R/h = 400$ and $v = 0.3$. Solutions were obtained to within engineering accuracy taking eight terms for both unknown functions $a_m(t)$ and $b_n(t)$. First, the unperturbed motion was found to have resonant responses at the axisymmetric natural frequencies, $\bar{\omega}_i$, where $i = 1, 3, 5, \ldots$, which are 35·2, 37·2, 46·4..., respectively, for the case when $\bar{q}_0 = 0$. Further, the shell is found to buckle at $q_{cr} = 63\cdot6\pi^2 D/L^2$ with the circumferential wave number $N_c = 18$, the axial waveform being antisymmetric with respect to the central section of the shell.

Relations between the natural frequencies, $\bar{\omega}_i$, and wave number, N, are illustrated in Fig. 4, with the effect of static load $\bar{q}_0 = q_0/q_{cr}$ taken into consideration. In the figure, N_1 and N_c indicate the wave numbers corresponding to the minimum natural frequency and the buckling load, respectively. It will be seen that the effect of the static load is most pronounced in the vicinity of the lowest natural frequencies of each order.

Based on the natural vibration characteristics thus obtained, the instability regions of both principal and combination resonances were determined using the procedure described above. The results for $\bar{q}_0 = 0$ are illustrated in Fig. 5, where the shaded areas correspond to the instability

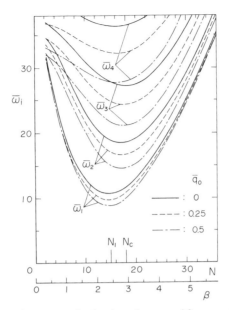

FIG. 4.   Effect of static compressive load on the natural frequencies of the clamped cylindrical shell: $Z = 100$, $R/h = 400$, $v = 0.3$.

regions. More details of some of these are given in Table 1. It should be noted that numerous instability regions (more than 70) exist in the limited frequency range here considered, which overlap with each other to form a continuous instability zone for some frequency range with the excitation amplitude q_1 exceeding only 1 % of the critical one, q_{cr}. Further, it will be seen that the openness of the instability region, θ_{ij}, becomes quite large

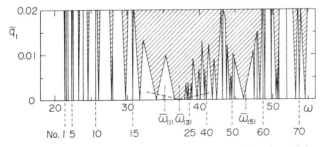

FIG. 5.   Instability regions of the clamped cylindrical shell under axial excitation: $Z = 100$, $R/h = 400$, $v = 0.3$, $\bar{q}_0 = 0$.

TABLE 1

INSTABILITY REGIONS OF THE CLAMPED CYLINDRICAL SHELL UNDER PULSATING COMPRESSIVE LOAD: $Z = 100$, $R/h = 400$, $v = 0.3$

No.	ω_{ij}	θ_{ij}	N	(i,j)	No.	ω_{ij}	θ_{ij}	N	(i,j)
1	21·26	0·010	15	1,1	21	37·24	(66·0)	18	2,2
2	21·30	0·028	14	1,1	22	37·32	8·78	16	2,2
3	21·60	0·018	16	1,1	23	37·70	1·40	19	2,2
4	21·73	0·036	13	1,1	24	37·88	1·05	15	2,2
5	22·27	0·057	17	1,1	25	38·24	0·599	16	1,3
⋮	⋮	⋮	⋮	⋮	26	38·38	0·476	15	1,3
15	30·71	0·051	8	1,1	27	38·41	0·536	17	1,3
16	31·78	1·16	23	1,1	28	38·45	1·37	20	2,2
17	34·05	3·94	24	1,1	29	38·48	0·078	6	1,1
18	34·20	0·474	7	1,1	30	38·80	0·775	14	2,2
19	36·46	2·92	25	1,1	⋮	⋮	⋮	⋮	⋮
20	37·11	(80·0)	17	2,2					

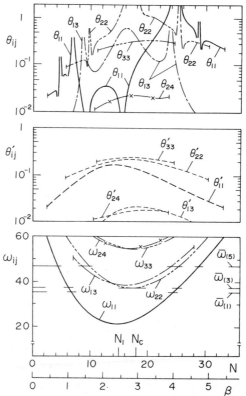

FIG. 6. Central frequency and the width parameter of the instability regions: $Z = 100$, $R/h = 400$, $v = 0.3$, $\bar{q}_0 = 0$.

when the central frequency, ω_{ij}, happens to be coincident with $\bar{\omega}_k$, where $k = 1, 3, 5 \ldots$, the axisymmetric natural frequencies of the unperturbed motion.

To make clear the global nature of the instability regions, values of ω_{ij} and θ_{ij} are plotted against the wave number N, with the results as shown in Fig. 6. In the figure, θ'_{ij} denotes the relative width parameter determined by assuming the unperturbed motion to be membrane state, that is, $N_{x0} = -(q_0 + q_1 \cos \Omega t)$, $N_{y0} = 0$ and $W_0 = \nu R(N_{x0}/Eh)$. The instability regions with θ_{ij} or θ'_{ij} less than 10^{-2} were omitted as they are of less importance. From the figure, one can easily find the central frequency, the instability

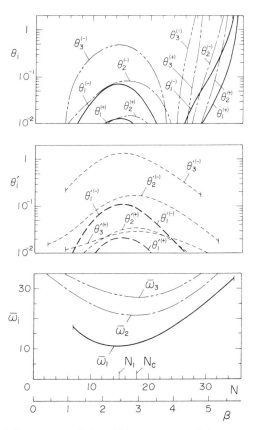

FIG. 7.   Central frequency and the width parameter of the secondary instability regions: $Z = 100$, $R/h = 400$, $\nu = 0.3$, $\bar{q}_0 = 0$.

width as well as the associated axial modes of vibration parametrically excited, for each wave number N. It is to be noted that the effect of pre-instability bending vibration on the width parameter θ_{ij} is quite significant, causing several spikes at the resonances of the unperturbed motion, where $\omega_{ij} \simeq \bar{\omega}_{(k)}$, with $k = 1, 3, 5 \ldots$.

The central frequency and the width parameter of the secondary instability regions are shown in Fig. 7, together with the width parameter θ_i'

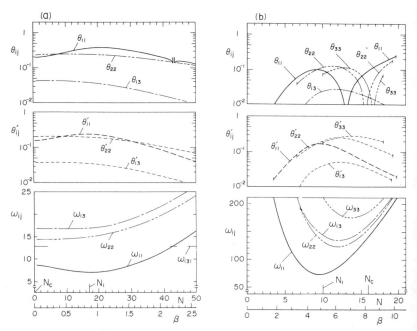

FIG. 8.   Central frequency and the width parameter of the instability regions: (a) $Z = 10$, $R/h = 400$, $\bar{q}_0 = 0$; (b) $Z = 1000$, $R/h = 400$, $\bar{q}_0 = 0$.

obtained by the membrane assumption. Although the width parameter θ_i increases significantly near the resonance of the unperturbed motion, the actual instability width for small excitation amplitude $\bar{q}_1$ will be smaller than those given above, as it is proportional to $(\bar{q}_1)^2$. Further, the results for the main instability regions for the cases when $Z = 10$ and 1000 are shown in Figs 8(a) and 8(b), respectively. It is to be noted that the width parameters θ_{ij} in these cases are not much different from θ_{ij}', the values obtained under the membrane assumption.

Finally, the following conclusions were obtained concerning the parametric instability of cylindrical shells under axial excitation:

(1) For each circumferential wave number, N, there exists the secondary and the principal instability regions at $\omega \simeq \bar{\omega}_i$ and $\omega \simeq 2\bar{\omega}_i$, respectively, as well as those of the combination resonance type at $\omega \simeq \bar{\omega}_i + \bar{\omega}_j$, where $i + j =$ even number.

(2) Judging from the magnitude of the relative width parameter, the principal instability regions with N around N_1 and N_c are generally most important while the secondary instability regions are of much less practical importance.

(3) The effect of the unperturbed bending vibration is quite significant for shells with Z ranging from 50 to 300, where the central frequencies of the main instability regions almost overlap with the range of the axisymmetric natural frequencies, causing resonant responses in the unperturbed motion. For shells with $Z \ll 50$ or $Z \gg 300$, the effect mentioned above becomes less significant and the membrane assumption will be justified.

(4) The static compressive load is found to have the effect of lowering the frequencies and enlarging the width parameters of the instability regions.

In the corresponding experimental studies, which will be described in the next section, the cylindrical shell almost filled with water is employed due to the insufficient capacity of the shaker. In this connection, it is to be added that both theoretical and experimental studies were conducted by Kana and Craig[48] on the dynamic stability of a liquid-filled cylinder under longitudinal base excitation while the effect of the surrounding fluid on the parametric excitation of an infinitely long cylindrical shell was studied by Deng and Popelar.[49] Only the linear problem for the determination of instability boundaries has been treated in this section. To clarify the shell response after the stability is lost, the corresponding non-linear problem should be considered, for which we cite an excellent review article presented by Hsu.[50]

5. EXPERIMENTAL RESULTS

In this section the results of preliminary experiments carried out to supplement the preceding theoretical analyses will be explained.

5.1. Test Cylinder and Test Set-Up

A test specimen with radius $R = 100$ mm, thickness $h = 0.247$ mm, length $L = 226.0$ mm was made of a commercial polyester film of nominal thickness, 0.25 mm, by lap-jointing along the longitudinal seam and attaching 11 mm thick duralumin end plates along both edges. The material properties, i.e. Young's modulus, E, Poisson's ratio, ν, and the mass density, ρ, were found to be 5.55 GPa, 0.30 and 1.405×10^3 kg m^{-3}, respectively, with which the geometric parameter $Z = \sqrt{1 - \nu^2}\, L^2/Rh$ becomes 1973.

The vibration exciter system of the test set-up is composed of the exciter control (Brüel & Kjaer type 1047, abbreviation BK 1047), power amplifier (BK 2706), vibration exciter (BK 4809), accelerometer (BK 4339), force transducer (BK 8201) and the conditioning amplifier (BK 2626), with which the sinusoidal sweeping excitation is possible for a wide range of frequencies keeping the amplitude of the exciter head or the shell response at the specific point constant.

The shell response was observed with a non-contact fibre-optics device, referred to as an 'MTI Model KD-320, Fotonic Sensor'. It has two measuring capabilities depending on the distance between the probe and the surface. It was found that when the maximum output voltage is adjusted to be 3.75 V and the probe is set at 70 μ from the shell surface, the sensitivity is 12.4 mV/μ in the range ± 50 μ, while when set at 1.60 mm from the surface, it is 0.71 mV/μ in the range ± 1.00 mm. The probe was mounted to a specially designed stand so that it could be moved in both longitudinal and circumferential directions along the shell surface. The measuring amplifier (BK 2607) and the narrow band spectrum analyser (BK 2031) were used to obtain the r.m.s. value and the frequency spectra of the output signal, respectively. Further, a dual-beam synchroscope was used to monitor the waveforms while a X–Y recorder was used for recording the frequency response characteristics. A general view of the test set-up for the parametric excitation of the shell is shown in Fig. 9.

5.2. Linear Free Vibration

The free vibration characteristics of the shell were determined by measuring the resonant responses of the shell when subjected to a lateral excitation at the point $x/L = 0.2$ and $\theta = 0°$, where x and θ are the axial and circumferential coordinates, respectively, and the longitudinal seam is located along $\theta = 354°$. The excitation amplitude ΔW was kept constant at 1.5 μ. A typical frequency response curve taken when $x/L = 0.5$ and $\theta = 187°$ is shown in Fig. 10, where $\Omega/2\pi$ is the excitation frequency in

FIG. 9.   General view of the test set-up for the parametric excitation of a circular cylindrical shell.

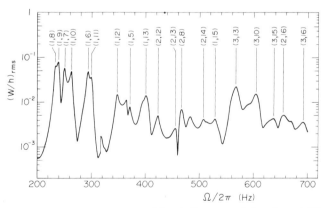

FIG. 10.   Frequency–response curve of the shell at $x/L = 0.5$ and $\theta = 187°$: excited at $x/L = 0.2$ and $\theta = 0°$ with $\Delta W = 1.5\,\mu$.

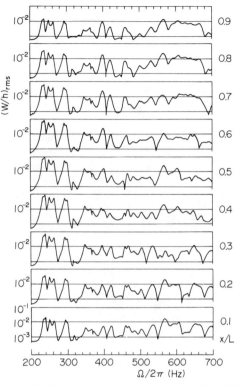

FIG. 11.   Variations of the frequency–response curves in the axial direction along
$\theta = 187°$.

Hertz while $(W/h)_{rms}$ means the r.m.s. value of the response W normalised
with the shell thickness h. The two numerals in parentheses indicate the
axial mode, i.e. the number of half-waves in the axial direction and the
circumferential wave number, respectively. The axial modes were determined from the set of frequency response curves as shown in Fig. 11,
which were recorded at various axial positions along $\theta = 187°$, while the
circumferential wave numbers were decided in a similar way from the
frequency response curves recorded with $3°$ intervals in θ. Some of the
vibration modes were determined more directly by recording the axial and
circumferential distributions of the r.m.s. value of the response as shown in
Fig. 12.

Natural frequencies thus determined are plotted against the wave
number N in Fig. 13, where m denotes the axial mode. Theoretical results,

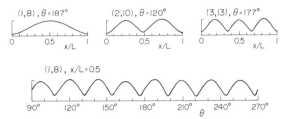

FIG. 12.   Axial and circumferential variations of the r.m.s. value of the responses for typical modes of vibration.

estimated from the graphs in Figs 2 and 3 for the case when $Z = 1973$ are shown, together with solid lines. It will be seen that both theory and experiment are in good agreement, demonstrating the practical applicability of the theoretical results presented in Section 2. It is to be added that the steady-state responses of a simply supported cylindrical shell under a concentrated lateral excitation have been analysed by Warburton.[51]

5.3. Non-Linear Vibration
To see the effect of amplitude on the natural vibration, the frequency

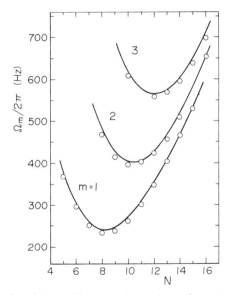

FIG. 13.   Comparison between theory and experiment for natural frequencies of a clamped cylindrical shell with $Z = 1973$.

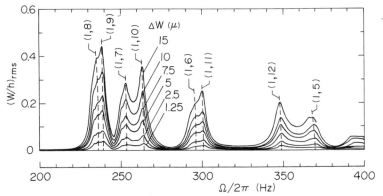

FIG. 14. Frequency–response curves of the shell at $x/L = 0.5$ and $\theta = 187°$: excited at $x/L = 0.2$ and $\theta = 0°$ with ΔW ranging from $1.25\,\mu$ to $15\,\mu$.

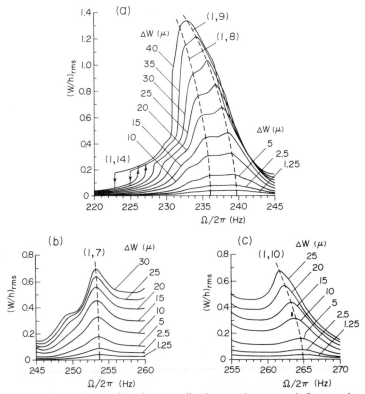

FIG. 15. Effect of the vibration amplitude on the natural frequencies: (a) vibrations with the modes (1, 8) and (1, 9); (b) vibration with the mode (1, 7); (c) vibration with the mode (1, 10).

response curves were obtained at $x/L = 0.5$ and $\theta = 187°$, with the excitation amplitude ΔW stepwisely increasing from $1.25\,\mu$ to $15\,\mu$. The results are shown in Fig. 14, where the dashed lines represent the so-called back-bone curves. The enlarged diagrams for the lowest four modes of vibration are illustrated in Figs 15(a), (b) and (c), where the angular position of the probe has been adjusted slightly to trace the peak response under each excitation amplitude. From these figures, a weak non-linear effect of softening type will be seen. More precisely, the natural frequencies for the lowest mode vibrations with $N = 7, 8, 9, 10$ decrease by about 0.6, 0.7, 0.9 and 1.3%, respectively, when $(W/h)_{rms} = 0.7$. In Fig. 15(a), the hysteresis loops with jump phenomena were found to be caused by the second order super-harmonic resonance of the natural vibration with the mode $(1, 14)$. No other significant non-linear phenomena were observed in the range of experiments here conducted, where the vibration amplitudes are limited to those shown in the figures due to the small capacity of the exciter with a rated force of 44.5 N.

5.4. Parametric Excitation

Owing to the insufficient capacity of the vibration exciter, the parametric excitation of the test cylinder was found possible only when the natural frequencies were considerably lowered by filling the cylinder with water. Hence, tests were conducted with the shell filled with water up to $x/L = 0.98$. First, the free vibration characteristics were determined with the same procedure as described in Section 5.2. A typical frequency response curve obtained at the central section of the shell is shown in Fig. 16, while the experimental results for the natural frequencies together

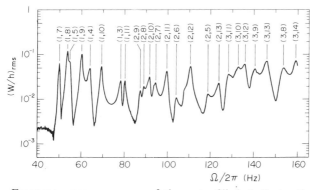

FIG. 16. Frequency–response curve of the water-filled shell at $x/L = 0.5$ and $\theta = 187°$: excited at $x/L = 0.2$ and $\theta = 0°$ with $\Delta W = 2.5\,\mu$.

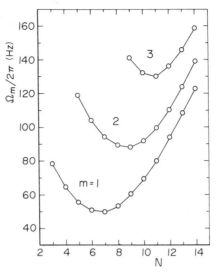

FIG. 17. Experimental results for the natural frequencies of the water-filled cylindrical shell with $Z = 1973$.

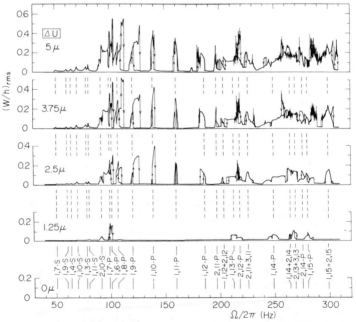

FIG. 18. Frequency–response characteristics of the parametrically excited water-filled cylinder at $x/L = 0.5$ and $\theta = 139°$: static compressive load is 16.2 N and ΔU ranges from 1.25 to $5\,\mu$.

with the corresponding modes are illustrated in Fig. 17. Comparing with Fig. 13, it can be seen that the natural frequencies are reduced to less than a quarter of the original ones due to the contained water.

The parametric excitation tests were conducted with the test set-up as shown in Fig. 9. A pulsating axial load and a constant static compressive load of 16·2 N were applied to the centre of the upper cover plate through the force transducer. The excitation frequency, $\Omega/2\pi$, was varied between 40 and 310 Hz at a rate of $0·2\,\mathrm{Hz\,s^{-1}}$, while the axial displacement amplitude ΔU at the upper end of the shell was kept constant to the prescribed levels, ranging from 1·25 to 5 μ. Meanwhile, the frequency responses at $x/L = 0·5$ and $\theta = 139°$ were recorded with the results as shown in Fig. 18. In the figure, 1,7-S and 1,7-P, for example, indicate the secondary and primary instability regions, respectively, in which the vibration with the mode (1,7) is excited, while 1,12 + 2,12, for instance, stands for the combination type instability region where the vibrations with the modes (1,12) and (2,12) are simultaneously excited. Classifications of the instability regions were made through the determination of the mode of the response as well as the frequency relations between excitation and response. It will be seen that the width of each instability region increases with the increase in the excitation amplitude ΔU. It is to be added that $\Delta U = 5\,\mu$ corresponds to 1·6 % of the critical compressive displacement, U_{cr}, of the test cylinder without water, which is predicted to buckle at 92·5 % of the classical buckling load P_{cl}.[52]

To ease the precise observation, enlarged diagrams for typical responses corresponding to each kind of instability region are shown in Figs 19(a), (b) and (c), respectively. The sweep rate in these cases was set at $0·1\,\mathrm{Hz\,s^{-1}}$, i.e. half of the one used above. The frequency spectra corresponding to these responses are also given in Fig. 20, where the location of the excitation frequency is marked with an inverse triangle. The frequency relations characterising the secondary, primary as well as the combination resonances will be clearly recognised.

From Figs. 17 through 20, it can be concluded that, in general, the results here obtained compare favourably with those theoretically predicted in Section 4 for the empty cylindrical shell under similar axial excitation. A few different points should be noted. The first is that the combination resonances associated with the two axial modes of vibration, i and j, with $i + j \neq$ even number, become possible in the experiments as will be seen in Fig. 18. This may be attributed to the loss of symmetry in the axial mode of vibration due to the contained water. The second point is that the effect of the unperturbed bending motion becomes less significant, which seems to

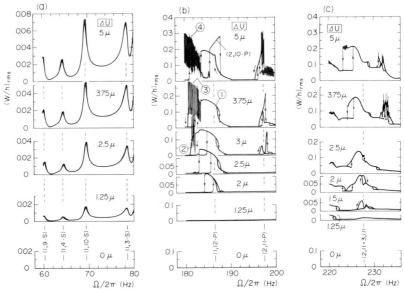

FIG. 19. Typical responses at various instability regions. (a) Secondary instability region; (b) principal instability region; (c) combination instability region.

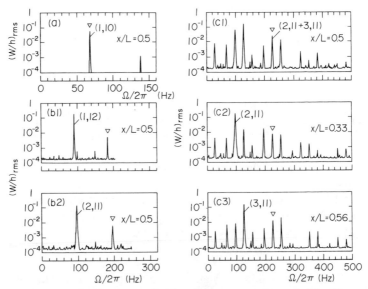

FIG. 20. Frequency spectra for the responses shown in Fig. 19.

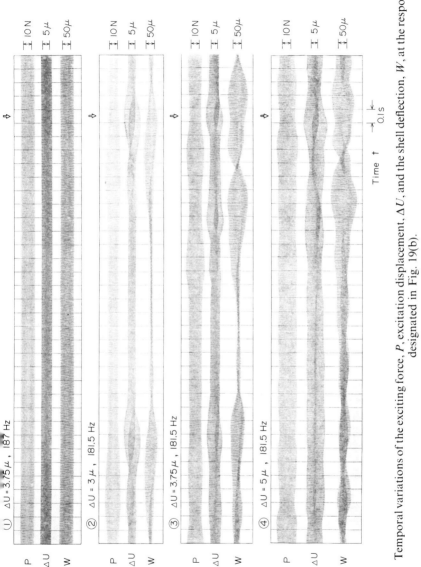

FIG. 21.   Temporal variations of the exciting force, P, excitation displacement, ΔU, and the shell deflection, W, at the responses designated in Fig. 19(b).

be attributed to the fact that the axial displacement, not the axial load, is controlled in the experiment, preventing the pre-instability resonant responses.

Finally, it is to be added that the travelling wave phenomena were observed to occur with the increase in the excitation amplitude and with the reduction in the sweep rate. Most of the dense recordings of the response with $(W/h)_{rms}$ ranging from 0·1 to 0·3 in Figs 18 and 19 correspond to these phenomena. For reference, temporal variations of the exciting force, P, excitation amplitude, ΔU and the shell bending displacement, W, are shown in Fig. 21, for the responses marked 1–4 in Fig. 19(b). The enlarged

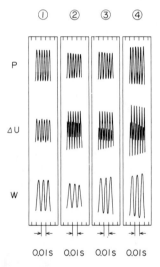

FIG. 22.   Enlarged diagrams for the portions of the responses indicated by arrows in Fig. 21.

diagrams for the portions of the responses indicated by arrows are shown in Fig. 22. The results for 1 stand for the steady-state response at the principal instability region while those in 2–4 represent the bifurcation as well as the consequent development of the travelling wave phenomena with the increase in the excitation amplitude, ΔU. It may be concluded that many more theoretical and experimental studies are required to clarify the whole aspect of the parametric excitation problem including these phenomena.

REFERENCES

1. FLÜGGE, W., *Ing.-Arch.*, **3**, 463–506, 1932.
2. FORSHERG, K., *AIAA J.*, **2**, 2150–7, 1964.
3. WARBURTON, G. B., *J. Mech. Engng. Sci.*, **7**, 399–407, 1965.
4. FORSBERG, K., *AIAA J.*, **7**, 221–7, 1969.
5. VRONAY, D. F. and SMITH, B. L., *AIAA J.*, **8**, 601–3, 1970.
6. WARBURTON, G. B. and HIGGS, J., *J. Sound Vib.*, **11**, 335–8, 1970.
7. GOLDMAN, R. L., *AIAA J.*, **12**, 1755–6, 1974.
8. SANDERS, JR, J. L., An improved first-approximation theory for thin shells, *NACA Tech. Rept. R-24*, 1960.
9. DYM, C. L., *J. Sound Vib.*, **29**, 189–205, 1973.
10. SHARMA, C. B., *J. Sound Vib.*, **35**, 55–76, 1974.
11. SHARMA, C. B., *J. Sound Vib.*, **63**, 581–92, 1979.
12. SOEDEL, W., *J. Sound Vib.*, **70**, 309–17, 1980.
13. DONNELL, L. H., Stability of thin-walled tubes under torsion, *NACA Tech. Rept. 479*, 1932.
14. YAMAKI, N., *Rept. Institute of High Speed Mech.*, *Tohoku Univ.*, **22**, 109–36, 1970.
15. ARMENÀKAS, A. E., *Proc. ASCE*, **EM5**, 95–109, 1967.
16. REISSNER, E., Non-linear effects in the vibrations of cylindrical shells, *Ramo-Wooldridge Corporation Report AM5-6*, 1955.
17. EVENSEN, D. A., in: *Thin-Shell Structures*, Y. C. Fung and E. E. Sechlev, Eds., Prentice-Hall, Englewood Cliffs, New Jersey, 1974, pp. 133–55.
18. ATLURI, S., *Int. J. Solids Structures*, **8**, 549–69, 1972.
19. KOITER, W. T., in: *Non-linear Problems*, R. E. Langer, Ed., University of Wisconsin Press, Madison, 1963, pp. 257–75.
20. BUDIANSKY, B., in: *Theory of Thin Shells*, F. I. Niordson, Ed., Springer-Verlag, Berlin, 1969, pp. 212–33.
21. REHFIELD, L. W., *Int. J. Solids Structures*, **9**, 581–90, 1973.
22. REHFIELD, L. W., *AIAA J.*, **12**, 388–90, 1974.
23. SANDERS, JR, J. L., *Q. Appl. Math.*, **21**, 21–36, 1963.
24. RADWAN, H. R. and GENIN, J., *Int. J. Non-Linear Mech.*, **10**, 15–29, 1975.
25. RADWAN, H. R. and GENIN, J., *J. Appl. Mech.*, *Trans. ASME*, **E-43**, 370–72, 1976.
26. HARARI, A., *Int. J. Non-Linear Mech.*, **11**, 169–81, 1976.
27. RAJU, K. K. and RAO, G. V., *J. Sound Vibr.*, **44**, 327–33, 1976.
28. UEDA, T., *J. Sound Vibr.*, **64**, 85–95, 1979.
29. YAMAKI, N. and MORI, A., *J. Sound Vibr.*, **71**, 336–46, 347–60, 1980.
30. YAMAKI, N., OTOMO, K. and CHIBA, M., *J. Sound Vibr.*, **79**, 23–42, 43–59, 1981.
31. YAMAKI, N. and KODAMA, S., *Int. J. Non-Linear Mech.*, **11**, 99–111, 1976.
32. YAMAKI, N. and MATSUDA, K., *Ing.-Arch.*, **45**, 79–89, 1976.
33. YAMAKI, N., OTOMO, K. and MATSUDA, K., *Exp. Mech.*, **15**, 23–8, 1975.
34. YAMAKI, N., in: *Buckling of Structures*, B. Budiansky, Ed., Springer-Verlag, Berlin, 1976, pp. 312–30.
35. YAMAKI, N., in: *Theoretical and Applied Mechanics*, W. T. Koiter, Ed., North-Holland, Amsterdam, 1977, pp. 461–76.

36. YAMAKI, N. and KODAMA, S., in: *Theory of Shells*, W.T. Koiter and G.K. Mikhailov, Eds., North-Holland, Amsterdam, 1980, pp. 635–67.
37. BOLOTIN, V. V., *The Dynamic Stability of Elastic Systems*, Holden-Day, London, 1964, p. 451.
38. YAO, J. C., *AIAA J.*, **1**, 1391–6, 1963.
39. YAO, J. C., *J. Appl. Mech.*, **E32**, 109–15, 1965.
40. WOOD, J. D. and KOVAL, L. R., *AIAA J.*, **1**, 2576–82, 1963.
41. VIJAYARAGHAVAN, A. and EVAN-IWANOWSKI, R. M., *J. Appl. Mech.*, *Trans.* *ASME*, **E34**, 985–90, 1967.
42. KALNINS, A., *J. Appl. Mech.*, *Trans. ASME*, **E41**, 1063–8, 1974.
43. KOVAL, L. R., *J. Acoust. Soc. Amer.*, **55**, 91–7, 1974.
44. HSU, C. S., in: *Thin Shell Structures*, Y.C. Fung and E.E. Sechler, Eds., Prentice-Hall, Englewood Cliffs, New Jersey, 1974, pp. 103–31.
45. YAMAKI, N. and NAGAI, K., *J. Sound Vibr.*, **45**, 513–27, 1976.
46. NAGAI, K. and YAMAKI, N., *J. Sound Vibr.*, **58**, 425–41, 1978.
47. HSU, C. S., *J. Appl. Mech.*, *Trans. ASME*, **E30**, 367–72, 1963.
48. KANA, D. D. and CRAIG, R. R., JR, *J. Spacecraft Rockets*, **5**, 13–21, 1968.
49. DENG, Z. F. and POPELAR, C. H., *J. Acoust. Soc. Amer.*, **52**, 1430–6, 1972.
50. HSU, C. S., in: *Advances in Applied Mechanics*, vol 17, R. von Mises and Th. von Karman, Eds., Academic Press, New York, 1977, pp. 245–301.
51. WARBURTON, G. B., *J. Engng. Industry, Trans. ASME*, **B96**, 994–9, 1974.
52. YAMAKI, N. and KODAMA, S., *Rep. Inst. High Speed Mech., Tohoku Univ.*, **25**, 99–141, 1972.

Chapter 4

EFFECTIVE WIDTHS IN PLATE BUCKLING

J. RHODES

Department of Mechanics of Materials,
University of Strathclyde, Glasgow, UK

SUMMARY

In this chapter the history of plate post-buckling analysis is briefly reviewed. Expressions are given for the effective widths of plates under various boundary conditions at loads near buckling. The behaviour of plates at loads far in excess of buckling is examined and the effects of change in wavelength are studied. Various empirical effective width expressions are described and the treatment of buckled plates in design codes is outlined. A form of 'effective width' expression is suggested for design use which is applicable at all stages of load or to all plate width to thickness ratios, and this is compared with existing design formulae.

1. BACKGROUND TO PLATE POST-BUCKLING ANALYSIS

The term 'effective width' has been used for a great many years in the description of plate post-buckling behaviour. Before any rigorous examinations of compressed plate behaviour had been carried out, it was recognised in ship design that thin plating was not fully effective in withstanding compressive loads. This was allowed for in design by assuming that a width of plate not greater than 25 times the plate thickness on each side of a stiffener could be considered to effectively withstand compression. Therefore for a wide plate with stiffeners on both unloaded edges it was assumed that a width of this plate equal to 50 times its thickness was effective.

In 1930 a large series of compression tests on plates of various materials was carried out by Schuman and Back.[1] Their results indicated that, indeed, for plates wide and thin enough to buckle under load the ultimate load which could be carried did not increase in proportion to the width, and was in fact relatively insensitive to increase in width. They concluded that after buckling a plate behaves as though only part of its width is effective in carrying load.

This phenomenon was investigated theoretically in 1932 by von Karman et al.[2] who obtained the first effective width expression. The expression derived was later shown to be equivalent to

$$\frac{b_e}{b} = \sqrt{\frac{\sigma_{CR}}{\sigma_Y}} \qquad (1)$$

where b is the plate width, b_e its effective width, σ_{CR} the plate buckling stress and σ_Y the material yield stress. For mild steel with commonly used properties, i.e. Young's modulus $E = 200 \times 10^3\,\text{N mm}^{-2}$, $\sigma_Y = 250\,\text{N mm}^{-2}$ and Poisson's ratio $v = 0.3$, the effective width at failure obtained from eqn (1) for a simply supported plate is equal to $53.8t$, where t is the plate thickness regardless of the actual plate width.

In the 1930s and 1940s many researchers tackled the problems of plate post-buckling strength and stiffness. Notable advances in the understanding of the problem were made by Cox,[3,4] Marguerre,[5] Levy,[6] Koiter,[7] Hemp,[8] van der Neut[9] and others. The major area of application was in the field of aircraft structures and in this period a great deal of light was shed on the behaviour of very thin plates, particularly those with boundary conditions applicable to aircraft structures. In the mid 1940s the first investigation of the effects of imperfections was made by Hu et al.[10]

At this time also, a second area of application of plate post-buckling analysis was becoming increasingly important, that is the field of cold-formed sections. The major development work was carried out in the USA by Winter[11] who produced, on the basis of experimental evidence, a modification to the effective width expression of von Karman to take account of imperfections, etc. This was incorporated into the AISI standard for cold-formed sections and has remained there, with minor modifications, to the present day. This effective width expression has also been adopted in the design specifications of a number of countries. It is of note that the boundary conditions of greatest importance in cold-formed section analysis are somewhat different to those of major importance in the aircraft structures field so that some differences in plate behaviour are to be expected.

Research into cold-formed sections in the UK began a little later than in the USA, and followed a slightly different approach with regard to plate behaviour. Some of the major researchers in the early days were Kenedi *et al.*,[12] Chilver[13,14] and Harvey.[15] The empirical treatment of local buckling obtained by Chilver from experiments on cold-formed sections formed the basis of the earlier editions of the British Standard specification for the use of cold-formed steel sections in building (BS 449 Addendum No. 1[23]).

During the 1950s and 60s further advances in the analysis of plate post-buckling behaviour were made by Stowell,[16] Coan,[17] Botman and Besseling,[18] Stein[19] and Yamaki[20] among others. Stowell examined the behaviour of plates with one unloaded edge free and the other edge simply supported, and took plastic behaviour into account. Botman and Besseling showed from the results of tests that elastically derived effective widths could be applied with accuracy to the analysis of aluminium plates in the plastic range. Coan, Stein and Yamaki all produced further illumination to the problems of plates loaded well beyond the buckling load with various types of boundary conditions and imperfections. Yamaki's research in particular produced results for a wide variety of boundary conditions.

In the late 1960s Graves Smith[21] applied rigorous plasticity theory to the analysis of plates. This paper was the forerunner to a very large number of elasto-plastic plate analyses of later years.

In 1969, Walker[22] obtained explicit expressions for the behaviour of a simply-supported square plate at loads far beyond buckling. This work provided the basis for the treatment of compressed plates in the current edition of BS 449 Addendum No. 1.[23]

The last decade or so has seen an extensive input into the analysis of plate behaviour. Analysis of plates under a wide variety of loading and boundary conditions has been carried out by the author and co-workers.[24-6] The perturbation approach used by Walker[22] has been extended by Williams and Walker[27] to provide explicit expressions for the behaviour of a number of different plate loading and boundary conditions at loads far beyond buckling.

A tremendous growth in the importance of local buckling in heavy civil engineering structures, largely related to box girder bridges, has resulted in a massive research input from civil engineers. The fact that the plating used in box girder bridges is of relatively low width to thickness ratios in comparison to those studied by the majority of researchers in the past has led to a great increase in the importance of plastic buckling and imperfection effects. Fortunately as the need for rigorous elasto-plastic analysis arose the capability of carrying out such analysis grew in parallel, with the use of numerical methods allied to high capacity computers. This has resulted in the production of a large number of papers on elasto-plastic plate behaviour in recent years. Notable among the researchers involved are Moxham,[28] Frieze *et al.*[28] Crisfield,[30] Little[31] and Rogers and Dwight.[32]

To summarise, there exists at present a substantial amount of information on plate compressional behaviour, obtained initially from researchers in ship and aeronautical design, followed by those involved in cold-formed sections and, more recently, in the heavy civil engineering field. The results of these researches are incorporated in various design specifications.

There are some differences in the requirements of engineers in the different fields. For example, bridge designers require very accurate knowledge of the behaviour of relatively thick plates and are not very concerned with the far post-buckling range. In the cold-formed sections field, on the other hand, knowledge of the far post-buckling range is very important, although the possibility of utilising the inelastic capacity of sections and the increasing capability of cold-forming manufacture has led to keen interest in relatively thick plate behaviour in this field also.

In bridge, aircraft and ship design, it is usually considered that for uniaxially loaded plates the unloaded edges will remain straight after buckling due to the action of adjacent plates in the structure. In plate elements of cold-formed sections, it is generally more realistic to assume that the unloaded edges are free to wave in the plane of the plate.

While there are differences in the research interests in the different fields, there are also many areas of similarity, and there is a need for an overall look at the various aspects of design analysis applicable to different plate boundary conditions. In this chapter formulae governing the behaviour of compressed plates, both in the range immediately after buckling and in the far post-buckling range, are presented and discussed.

There have been some excellent reviews of plate post-buckling behaviour published in the past, notably those by Jombock and Clark,[33] Koiter[34] and Supple and Chilver.[35] While some of the expressions given in these papers are reproduced here, for reasons of space, many are omitted, and readers interested in this field should find these papers very informative.

2. THE EFFECTIVE WIDTH CONCEPT

In the analysis of plate post-buckling behaviour the term 'effective width' or 'equivalent width' has been applied to describe different effects. The same terms have also been applied to cover quite different aspects of plate behaviour such as shear lag, bending of wide flanges, etc. It is therefore necessary to define the term effective width in different forms according to the purpose required.

2.1. Effective Width for Strength

The form of membrane stress distribution usually encountered across the critical section (generally at the crest of a buckle) of a uniformly compressed plate, with both unloaded edges supported, is illustrated in Fig. 1(a). The maximum stress occurs at the plate edges while the stresses near the heavily buckled plate centre are relatively small. It is quite realistic in such a situation to consider that the effectiveness of the plate in withstanding load is confined to the plate edges. Under these circumstances the idealisation shown in Fig. 1(b) can be used, where the total load is carried by two strips of combined width b_e situated at the edges of the plate and carrying the maximum membrane stress sustained by the plate.

The effective width for strength, b_e, may therefore be defined as that width of fully effective (i.e. unbuckled) plate which sustains the same maximum membrane stress as the buckled plate under a given load.

Figure 1(c) shows a typical variation of σ_{AV}, the average membrane stress with σ_{max}, the maximum membrane stress, for a perfect plate. It can easily be shown that the ratio of effective width to full width at any point is the same as the ratio of σ_{AV} to σ_{max}, i.e.

$$\frac{b_e}{b} = \frac{\sigma_{AV}}{\sigma_{max}} \qquad (2)$$

It has been shown by a number of researchers in the past that the maximum

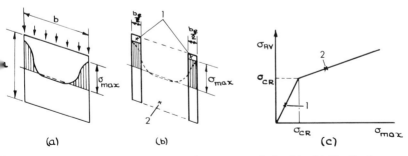

(a)   (b)   (c)

FIG. 1.   (a) Stress distribution across plate at crest of a buckle. (b) Idealised stress distribution (1. effective regions; 2. ineffective region). (c) Growth of maximum stress with average stress (1. $\partial\sigma_{AV}/\partial\sigma_{max} = 1$; 2. $\partial\sigma_{AV}/\partial\sigma_{max} < 1$).

load a plate can withstand is very close to that which causes first membrane yield to occur, so that an effective width based on the maximum membrane stress is useful in predicting the ultimate strength of a plate.

2.2. Effective Width for Stiffness

The average stress distribution across a buckled plate generally has similar characteristics to that of Fig. 1(a) with some redistribution of the stresses so that the average stress at the edges is not quite so high as the maximum stress of this figure. Since the edges of the plate do not buckle it can be shown that simple strength of materials rules relate the average edge stress to the average strain and the end displacement in the region of the edges. For a uniaxially compressed plate the relationships are

$$average\ edge\ stress \equiv \sigma_E = E\varepsilon_x = E\frac{u}{a} \tag{3}$$

where ε_x is the average strain in the x direction and u is the end displacement.

An effective width based on the average edge strain can be used to specify the compressional stiffness of a buckled plate. The effective width for stiffness, $\bar{b}_e$, can therefore be defined as that width of unbuckled plate which sustains the same average strain as the buckled plate for a given load.

Figure 2 shows the variation of σ_{AV} with ε_x for a uniaxially compressed plate. In such a case the ratio of $\bar{b}_e$ to b is the same as that of σ_{AV} to $E\varepsilon_x$ i.e.

$$\frac{\bar{b}_e}{b} = \frac{\sigma_{AV}}{E\varepsilon_x} \tag{4}$$

In the case of uniaxially compressed plates $\bar{b}_e$ is always greater than b_e.

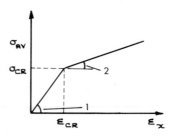

FIG. 2. Variation of average stress with strain for a perfect plate. 1. slope $\propto E$; 2 slope $\propto E^*$.

2.3. Reduced Tangent Modulus

From Fig. 2 the plate stiffness against compression is immediately reduced after local buckling. While this is taken into account by the effective width formulation of eqn (4) which gives the overall stiffness of the plate, it is sometimes important to know the magnitude of the instantaneous, or tangent stiffness, $\partial\sigma_{AV}/\partial\varepsilon_x$. The ratio of $\partial\sigma_{AV}/\partial\varepsilon_x$ after buckling to that before buckling has been termed the 'tangent effective width' by some researchers in the past, but is more commonly described using a reduced effective Young's modulus E^*. Using this formulation the ratio of post-buckling compressional stiffness to that before buckling is given by the ratio E^*/E.

3. EFFECTIVE WIDTH EXPRESSIONS AT LOADS NOT GREATLY IN EXCESS OF BUCKLING

In most of the commonly examined cases of plate behaviour the buckled form assumed by a perfect plate at the instant of buckling changes very gradually as load increases. At loads below about twice the buckling load the assumption of an unchanging buckled form gives results of engineering accuracy. As an indication of the range of plates which may be covered accurately by such an assumption, in the case of simply supported plates, the analysis of structural steel plates of width to thickness ratio of 120 or less can be accomplished with reasonable accuracy. Using this assumption permits the plate behaviour in the initial post-buckling range to be predicted by simple expressions of the form

$$\frac{b_e}{b} = \frac{\sigma_{AV}}{\sigma_{max}} = \frac{\partial\sigma_{AV}}{\partial\sigma_{max}} + \left(1 - \frac{\partial\sigma_{AV}}{\partial\sigma_{max}}\right)\frac{\sigma_{CR}}{\sigma_{max}} \tag{5}$$

$$\frac{\bar{b}_e}{b} = \frac{\sigma_{AV}}{E\varepsilon_x} = \frac{E^*}{E} + \left(1 - \frac{E^*}{E}\right)\frac{\varepsilon_{CR}}{\varepsilon_x} \tag{6}$$

where $\sigma_{CR} = (K\pi^2 D/b^2 t)$, D being the plate flexural rigidity factor $Et^3/12(1-v^2)$ and K the buckling coefficient, and $\varepsilon_{CR} = (\sigma_{CR}/E)$ for uniaxial compression.

4. VARIATION OF PLATE BEHAVIOUR WITH BUCKLE HALF WAVELENGTH

All coefficients in Table 1 have been based on specified values of e, the ratio of buckle half wavelength to plate width. The values of the coefficients are

TABLE 1
COEFFICIENTS IN EFFECTIVE WIDTH EXPRESSIONS

Plate loading	Aspect ratio a/b	$\dfrac{E^*}{E}$	$\dfrac{\partial \sigma_{AV}}{\partial \sigma_{max}}$	K	References[d]
SS(1) / SS(1) (a) → ←	1	0·408	0·26	4	4
SS(1),SS(2) / SS(2) (b) → ←	1	0·5	0·5	4	5
SS(2),SS(3) / SS(3) (c) → ←	—	0·746	—	—	33
SS(3),CL(1) / CL(1) → ←	$\begin{cases} 0·66 \\ 0·667 \end{cases}$	0·45 0·481	0·374 0·391	— 6·97	27 26
CL(1),CL(2) / CL(2) → ←	$\begin{cases} — \\ 0·66 \end{cases}$	0·497 0·462	— 0·452	6·97 —	34 27
CL(2),SS(1) / CL(1) → ←	1	0·494	0·29	5·739	26
SS(1) / F(1),CL(1) → ←	$\begin{cases} \text{any} \\ 2 \end{cases}$	4/9 0·438	4/9 0·365	— 0·668	16 26
CL(1) / F(1) → ←	2	0·556	0·485	1·335	26

[a] (1) Unloaded edges free from normal and shear stresses.
[b] (2) Unloaded edges constrained to remain straight with zero nett load.
[c] (3) Unloaded edges completely constrained from in-plane movement.
[d] Not necessarily reference of the original investigator.
SS Simple support conditions on plate edge.
CL Clamped at fixed support conditions.
F   Free support conditions.

TABLE 2

VARIATION OF PLATE COEFFICIENTS WITH BUCKLE HALF WAVELENGTH (e)

Plates with stress-free unloaded edges—loaded edges simply supported:

Unloaded edge condition	Buckling coefficient, K	$\dfrac{E^*}{E}$	$\dfrac{\partial \sigma_{AV}}{\partial \sigma_{max}}$
1. Simply supported (SS)	$\left(e + \dfrac{1}{e}\right)^2$	$1 - \dfrac{2}{\left(3 + 2 \cdot 151 e^3/(e^3 + 4 \cdot 659)\right)}$	$0 \cdot 25 + 0 \cdot 0045e + \dfrac{0 \cdot 0833}{(1 + 14e^4)}$
2. Fixed	$2 \cdot 49 + 5 \cdot 139e^2 + \dfrac{0 \cdot 975}{e^2}$	$1 - \dfrac{2}{\left(3 \cdot 383 + 7 \cdot 1e^3/(e^3 + 4 \cdot 616)\right)}$	$0 \cdot 35 + 0 \cdot 013e + \dfrac{0 \cdot 0488(1 + e)}{(1 + 5e^2)}$
3. One edge SS—one edge free	$\dfrac{0 \cdot 977}{e^2} + 0 \cdot 425$	$1 - \dfrac{2}{\left(3 + 1 \cdot 121e/(1 + 1 \cdot 531e)\right)}$	$\simeq 0 \cdot 365$ throughout range for $e > 0 \cdot 5$
4. One edge fixed—one edge free	$\dfrac{0 \cdot 934}{e^2} + 0 \cdot 128e^2 + 0 \cdot 59$	$1 - \dfrac{2}{\left(3 + 2 \cdot 328e/(1 + 1 \cdot 082e)\right)}$	$\simeq 0 \cdot 47$ throughout range for $e > 0 \cdot 7$

TABLE 2—contd.

Plates with all edges kept straight—all edges simply supported.

Loading conditions	Buckling coefficient, K	$\dfrac{E^*}{E}$	$\dfrac{\partial \sigma_{AV}}{\partial \sigma_{max}}$
5. Uniaxial compression, $\sigma_{yAV}=0$	$\left(e+\dfrac{1}{e}\right)^2$	$\dfrac{1+e^4}{3+e^4}$	$\dfrac{1+e^4}{3+e^4}$
6. Displacement in y direction prevented, $\varepsilon_y=0$	$\dfrac{\left(e+\dfrac{1}{e}\right)^2}{1+\nu e^2}$	$\dfrac{(1+e^4)(3e^4+1)(1-\nu^2)}{(3+e^4)(1+3e^4)-(2e^2-\nu(1+e^4))^2}$	$\dfrac{(1+3e^4)(1+e^4)}{(3+e^4)(1+3e^4)-2e^2(2e^2-\nu(1+e^4))}$
7. Biaxial loading, $\sigma_{yAV}=\eta\sigma_{xAV}$	$\dfrac{\left(e+\dfrac{1}{e}\right)^2}{1+e^2\eta}$	$\dfrac{(1+e^4)(1-\nu\eta)}{3+e^4+\eta(2e^2-\nu(1+e^4))}$	$\dfrac{1+e^4}{3+e^4+2\eta e^2}$

not constant with respect to e but only apply to the specific half wavelength for which they were calculated, which is in most cases that which produces the minimum buckling coefficient. In some cases this cannot be achieved in practice, particularly if a plate is short in comparison to its width. It is also worth mentioning that consideration of that buckle half wavelength which produces minimum K does not necessarily give conservative results for other half wavelengths after buckling. (This will be examined in greater detail later.) As the behaviour of a buckled plate can vary substantially with the buckle half wavelength, it is useful to be able to evaluate the coefficients for any specified half wavelength. Details of the variation of coefficients are given in Table 2.

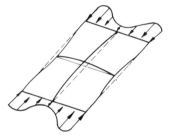

FIG. 3. Plate with unloaded edges stress free and able to wave.

In the case of plates which have unloaded edges free to wave (Fig. 3) the expressions given in the table are obtained on the basis of curve fitting to computer results of the writer for a large range of e values. The computer analysis is extremely accurate in the initial post-buckling range and it is expected that all values obtained using the expressions given should be very close to the exact values in the immediate post-buckling range for $0\cdot4 < e < 4$.

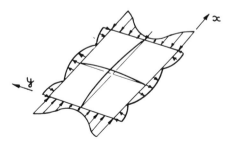

FIG. 4. Plate with all edges kept straight.

In the case of plates with straight edges on all sides (Fig. 4) rigorous analysis of the initial post-buckling behaviour can be carried out with e included as a variable. The behaviour of these plates has been analysed by many investigators since Marguerre[5] obtained the first rigorous solutions. The expressions given in Table 2 for these cases are particular cases obtained using the general expressions:

$$\varepsilon_x = \frac{1}{E(1 + e^4)} [\sigma_{xAV}(3 + e^4) + \sigma_{yAV}(2e^2 - v(1 + e^4)) + 2E\varepsilon_K] \quad (7)$$

$$\varepsilon_y = \frac{1}{E(1 + e^4)} [\sigma_{yAV}(1 + 3e^4) + \sigma_{xAV}(2e^2 - v(1 + e^4)) + 2Ee^2\varepsilon_K] \quad (8)$$

$$\sigma_{x\,max} = \frac{\sigma_{xAV}}{(1 + e^4)} \left[3 + e^4 + 2\frac{\sigma_{yAV}}{\sigma_{xAV}}e^2 \right] + \frac{2E\varepsilon_K}{(1 + e^4)} \quad (9)$$

where

$$\varepsilon_K = \left(e + \frac{1}{e} \right)^2 \frac{\pi^2 D}{b^2 tE} \quad (10)$$

In the above, the subscripts x and y refer to stresses and strains in the x and y directions, respectively.

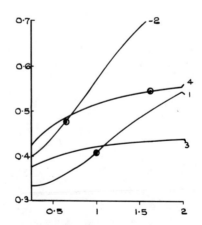

FIG. 5.   Variation of E^*/E (ordinate) with e (abscissa) for plates with stress free unloaded edges. $\odot \equiv e$ for K_{min}.

The variations of E^*/E and $\partial\sigma_{AV}/\partial\sigma_{max}$ are shown in Figs 5–8. Points of note for each case (Table 2) and as follows:

Case 1. $K_{min} = 4$ at $e = 1$, when $\dfrac{E^*}{E} = 0.408$ and $\dfrac{\partial\sigma_{AV}}{\partial\sigma_{max}} = 0.26$.

Case 2. $K_{min} = 6.967$ at $e = 0.66$ when $\dfrac{E^*}{E} = 0.474$ and $\dfrac{\partial\sigma_{AV}}{\partial\sigma_{max}} = 0.384$.

Case 3. $K_{min} \rightarrow 0.425$ as $e \rightarrow \infty$ when $\dfrac{E^*}{E} = 0.464$ and $\dfrac{\partial\sigma_{AV}}{\partial\sigma_{max}} \simeq 0.365$.

(Variation of E^*/E with e is small over most of the range.)

Case 4. $K_{min} = 1.282$ at $e = 1.644$, when $\dfrac{E^*}{E} = 0.543$ and $\dfrac{\partial\sigma_{AV}}{\partial\sigma_{max}} = 0.47$

Case 5. $K_{min} = 4$ at $e = 1$, when $\dfrac{E^*}{E} = \dfrac{\partial\sigma_{AV}}{\partial\sigma_{max}} = 0.5$

Case 6. $K_{min} = 2.8$ at $e = 1.581$, when $\dfrac{E^*}{E} = 0.746$ and $\dfrac{\partial\sigma_{AV}}{\partial\sigma_{max}} = 0.849$.

Case 7. Bulson[36] has shown that if η is positive and less than 0.5 then K has a minimum when $e = 1/\sqrt{1 - 2\eta}$. If η is negative (i.e. tensile loading on these sides) then this equation still holds, but if η is greater than 0.5 and positive then the buckling coefficient

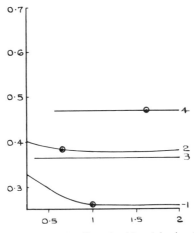

FIG. 6. Variation of $\partial\sigma_{AV}/\partial\sigma_{max}$ (ordinate) with e (abscissa) for plates with stress free unloaded edges. $\odot \equiv e$ for K_{min}.

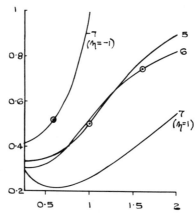

FIG. 7. Variation of E^*/E (ordinate) with e (abscissa) for plates with all edges kept straight. $\odot \equiv e$ for $K_{\min}$.

reduces indefinitely with increase in e and tends to a limiting value of $1/\eta$ as e becomes very large. For the particular case of equal loads in x and y directions on a square plate then $K = 2$, $E^*/E = 0.259$ and $\partial\sigma_{AV}/\partial\sigma_{x\max} = \frac{1}{3}$.

In general for plates with stress free edges $\partial\sigma_{AV}/\partial\sigma_{\max}$ is not very sensitive to variation in buckle half wavelength. The variation in E^*/E is more significant but not so substantial as is the case for plates whose edges are

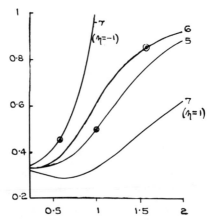

FIG. 8. Variation of $\partial\sigma_{AV}/\partial\sigma_{\max}$ (ordinate) with e (abscissa) for plates with all edges kept straight. $\odot \equiv e$ for $K_{\min}$.

kept straight. For such plates if the half wavelength is long enough then the effects of buckling will be negligible on plate stiffness or membrane stress levels, whereas it would appear that if the unloaded edges are stress free then buckling will have substantial effects even if the half wavelength could be artificially made very long. In the case of biaxially loaded plates the post-buckling behaviour is very sensitive to the loading in both directions, compressive loads on x and y directions causing much more severe conditions than combinations of tensile and compressive loads.

5. EFFECTS OF IMPERFECTIONS

The presence of imperfections, or initial deflections, is well known to seriously affect the behaviour of compressed plates, particularly at loads near the theoretical buckling load. It is generally considered that the most detrimental type of imperfection is one which has the same shape as the local buckle and this type of imperfection is, quite fortunately, relatively simple to analyse. In such a case the behaviour of the imperfect plate can be examined using simple modifications to the perfect plate analysis.

Unfortunately, the direct relationships between effective widths and loading can no longer be applied, as these must now be related via the deflection magnitudes w and imperfection magnitudes w_0. The governing expressions are as follows:

$$\frac{\varepsilon_x}{\varepsilon_{CR}} = \left(1 - \frac{w_0}{w}\right) + \frac{A}{K}\left[\left(\frac{w}{t}\right)^2 - \left(\frac{w_0}{t}\right)^2\right] \tag{11}$$

$$\frac{\sigma_{xAV}}{\sigma_{CR}} = \left(1 - \frac{w_0}{w}\right) + \frac{A}{K}\frac{E^*}{E}\left[\left(\frac{w}{t}\right)^2 - \left(\frac{w_0}{t}\right)^2\right] \tag{12}$$

$$\frac{\sigma_{x\,max}}{\sigma_{CR}} = \left(1 - \frac{w_0}{w}\right) + \frac{A}{K}\frac{E^*}{E}\left[\left(\frac{w}{t}\right)^2 - \left(\frac{w_0}{t}\right)^2\right]\bigg/\frac{\partial\sigma_{xAV}}{\partial\sigma_{x\,max}} \tag{13}$$

The values of K, E^*/E and $\partial\sigma_{xAV}/\partial\sigma_{x\,max}$ are those obtained in the perfect plate analysis, and if w_0 is set equal to zero, perfect plate behaviour is obtained.

To obtain the effective width variations for given imperfection amplitude w_0, the values of ε_x, σ_{xAV} and $\sigma_{x\,max}$ can be obtained for postulated values of w and the effective widths evaluated using $b_e/b = \sigma_{xAV}/\sigma_{x\,max}$ and

$\bar{b}_e/b = \sigma_{xAV}/E\varepsilon_x$. Expressions for the coefficients A for the cases examined in Table 2 are as follows:

Case 1. $A = \pi^2/[0{\cdot}6 + 1{\cdot}6e - 0{\cdot}1/(e - 0{\cdot}25)]^2$ (14)

Case 2. $A = 1/[0{\cdot}265 + 0{\cdot}364e - 0{\cdot}02/(e - 0{\cdot}25)]^2$ (15)

Case 3. $A = \dfrac{1}{0{\cdot}785e}$ (16)

Case 4. $A = \dfrac{1}{0{\cdot}785e}$ (17)

Cases 5, 6 and 7. They can be covered by the general expression

$$A = \frac{2{\cdot}73(1 + e^4)}{4e^2(1 + \eta e^2)\dfrac{E^*}{E}}$$ (18)

For the particular case of a square simply supported plate, $e = 1$, with stress free edges, i.e. case 1 in Table 2, the relevant coefficients are $E^*/E = 0{\cdot}408$, $\partial\sigma_{AV}/\partial\sigma_{max} = 0{\cdot}26$, $K = 4$ and $A = 2{\cdot}31$. Using these values gives the variation of effective widths in the post-buckling range for various imperfection magnitudes as shown in Figs 9 and 10. The imperfections cause merging of the pre-buckling and post-buckling paths. If they are small then the effective width curves are rounded off near the intersection of the pre-buckling and post-buckling ranges. If they are very large then the plate can be substantially ineffective from the start of loading.

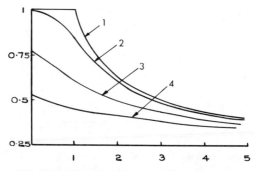

FIG. 9.   Effects of imperfections on effective width for strength. Ordinate: b_e/b. Abscissa: $\varepsilon_{max}/\varepsilon_{CR}$. Curve 1, $w_0/t = 0$; curve 2, $w_0/t = 0{\cdot}1$; curve 3, $w_0/t = 0{\cdot}5$; curve 4, $w_0/t = 1$.

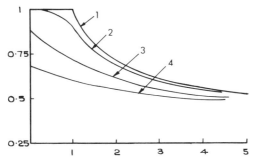

FIG. 10. Effects of imperfections of effective width for stiffness. Ordinate: b_e/b. Abscissa: $\varepsilon/\varepsilon_{CR}$. Curve 1, $w_0/t = 0$; curve 2, $w_0/t = 0{\cdot}1$; curve 3, $w_0/t = 0{\cdot}5$; curve 4, $w_0/t = 1$.

6. EFFECTIVE WIDTHS AT LOADS FAR IN EXCESS OF BUCKLING

At loads substantially greater than buckling the deflected form of a plate changes, the post-buckling stiffness reduces, and the predictions based on an unchanging deflected form can become highly inaccurate. A large number of researchers have obtained analyses of various problems in this range in the past, many of whom are named in the reference list. Because of the complexity of analysis involved the majority of results of these analyses have been presented in graphical form rather than in the form of explicit equations.

To overcome this difficulty a number of investigators have fitted mathematical expressions to their analytical curves, thus enabling the reader to make the best use of these results.

In recent years the perturbation method of analysis has been developed by Walker[22] and Williams and Walker[27] to obtain explicit expressions for a variety of plate problems at loads far beyond buckling. These permit direct evaluation of the relationships between σ_{max}, σ_{AV} and ε_x in the case of perfect plates. For imperfect plates an indirect method of evaluation, similar to that of the previous section must be employed.

From reference 22 the following expressions are available for square simply supported plates:

(a) Unloaded edges stress free:

$$\frac{\sigma_{max}}{\sigma_{CR}} = 2{\cdot}857\frac{\sigma_{AV}}{\sigma_{CR}} + 0{\cdot}486\left(\frac{\sigma_{AV}}{\sigma_{CR}}\right)^2 - 2{\cdot}343 \tag{19}$$

$$\frac{\varepsilon_x}{\varepsilon_{CR}} = \frac{\sigma_{AV}}{\sigma_{CR}} + 1{\cdot}45\left(\frac{\sigma_{AV}}{\sigma_{CR}} - 1\right) + 0{\cdot}294\left(\frac{\sigma_{AV}}{\sigma_{CR}} - 1\right)^2 \tag{20}$$

(b)   Unloaded edges kept straight:

$$\frac{\sigma_{max}}{\sigma_{CR}} = 1{\cdot}653\frac{\sigma_{AV}}{\sigma_{CR}} + 0{\cdot}174\left(\frac{\sigma_{AV}}{\sigma_{CR}}\right)^2 - 0{\cdot}827 \qquad (21)$$

$$\frac{\varepsilon_x}{\varepsilon_{CR}} = \frac{\sigma_{AV}}{\sigma_{CR}} + \left(\frac{\sigma_{AV}}{\sigma_{CR}} - 1\right) + \frac{5}{64}\left(\frac{\sigma_{AV}}{\sigma_{CR}} - 1\right)^2 \qquad (22)$$

Modifications to take imperfections into account are given in reference 22 but for simplicity the perfect case is considered here. As mentioned previously imperfection effects are greatest in the region of the buckling load and are not substantial at loads far in excess of buckling.

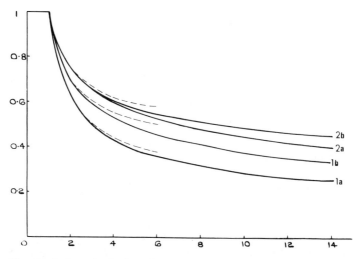

FIG. 11.   Variation of effective widths in the far post-buckling range. Ordinate: curves (a), b_e/b; curves (b), $\bar{b}_e/b$. Abscissa: curves (a), σ_{max}/σ_{CR}; curves (b), $\varepsilon/\varepsilon_{CR}$. Curves 1 are for stress free unloaded edges; curves 2 are for straight unloaded edges.

The effective widths obtained using these equations are shown in Fig. 11. Effective widths for plates with stress free edges are always less than those obtained if the unloaded edges are constrained to remain straight. In the case of plates with straight unloaded edges, differences arise between b_e and $\bar{b}_e$ as the plate is loaded substantially beyond buckling, although the differences are not so great as for plates with stress free unloaded edges. The dotted curves indicate the predictions based on non-changing deflected form and indicate that these are very accurate near the buckling load, and gradually lose accuracy as loading progresses.

6.1. Effects of Variation in Buckle Half Wavelength

Since simply supported plates naturally buckle into half wavelengths equal to their width, it could be postulated that the results obtained from an examination of square plates would be conservative (i.e. underestimate strength and stiffness) for plates with other buckle half wavelengths. This however is not the case. As shown previously, although the buckling coefficient for a simply supported plate has its minimum if the plate is square, the initial post-buckling stiffness reduces with reduction in half wavelength. Therefore a short plate $(a < b)$ may have a higher buckling load than a square plate but may have smaller effective widths at loads beyond buckling. For long plates, such as elements of thin-walled sections, the plate which initiates buckling is generally constrained by adjacent plates and the buckle half wavelength is less than the plate width.

There is also much evidence that there is a tendency for the buckle half wavelength to shorten after buckling. In a theoretically infinitely long plate this shortening can take place continuously. In a plate of finite length there are also mechanisms whereby this shortening can occur. Explosive changes in half wavelength have been observed by a number of experimenters, and this has been theoretically analysed by Cox,[4] Stein[19] and Supple[37] among others. Even without explosive or dynamic changes in buckle length the effects of changes in buckle wavelength can be induced by redistribution of the general buckle form along the plate so that at the critical or most highly loaded section of plate the buckle length is adjusted to make the most severe conditions possible apply. Moxham, for example, observed that in long simply supported plates loaded into the plastic range the buckle which finally becomes predominant has a half wave shorter than its width. There is evidence to show that in columns and beams the buckle half wavelengths of the elements can redistribute after local buckling to give the worst effects.

In order to evaluate the most adverse plate behaviour it is therefore necessary to consider the theoretically infinitely long plate in which the buckle half wavelength can change continuously. This was first investigated for plates with unloaded edges kept straight by Koiter.[7]

In the case of plates which have the unloaded edges kept straight the effects of wavelength on post-buckling stiffness are much greater than for plates with stress-free edges, as has been discussed previously, so it could be expected that for such plates changes in wavelength after buckling could be substantial. This can be examined very simply by neglecting any changes in deflected form across the plate and considering only shortening of the half wavelength. It is found that the half wavelength change at loads near buckling for this type of plate is approximately given by

$$e = \left(\frac{\varepsilon_{CR}}{\varepsilon_x}\right)^{1/3} \tag{23}$$

Substitution into the expressions for case 5 of Table 2 enables the post-buckling behaviour under these circumstances to be obtained. Figure 12 shows the variation of σ_{AV} with ε_x for such a plate (curve 3). Also shown are

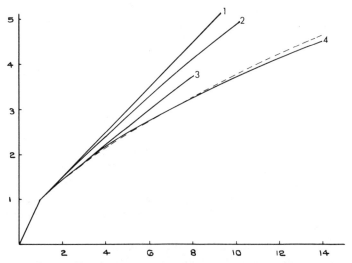

FIG. 12. Average stress-strain curves for simply supported plates, edges kept straight. Ordinate: σ_{AV}/σ_{CR}. Abscissa: $\varepsilon/\varepsilon_{CR}$. Curve 1: $e = 1$, buckled form unchanged; curve 2: $e = 1$, buckled form varies; curve 3: e varies, buckled form unchanged; curve 4: e varies, buckled form varies. (Dotted curve: $e = 1$, stress free edges.)

the corresponding curves for a plate with $e = 1$ computed for an unchanging buckle form (curve 1) and from eqn (22) (curve 2). The effects of changes in buckled form combined with changes in wavelength obtained using a simple approximate analysis outlined in reference[38] are shown in the figure (curve 4). At high strains Koiter[34] indicated that $\bar{b}_e/b$ tends to $0.785(\varepsilon_{CR}/\varepsilon_x)^{1/3}$. This is borne out by the analysis behind curve 4.

The figure indicates that change in wavelength has a substantial effect on plate behaviour. Comparing curves 1, 2 and 3 shows that in fact change in wavelength has more adverse effects than change in buckled form for this case.

It is interesting to realise that the minimum load curve 4 predicts that the plate is less stiff at high loads than a square plate with stress free unloaded edges, shown by the dotted curve.

Figure 13 shows the variation of $\bar{b}_e/b$ for square plates and for infinite plates. For square plates the curves obtained from eqns (20) and (22) are drawn and also that of Levy[6] who employed more degrees of freedom.

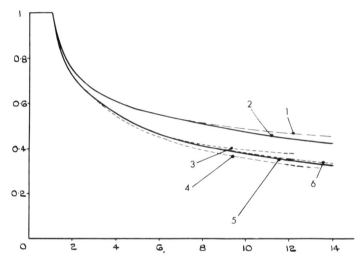

FIG. 13. Comparison of stiffness behaviour of square plates and infinite plates. Ordinate: b_e/b. Abscissa: $\varepsilon/\varepsilon_{CR}$. Curve 1: eqn (22); curve 2: straight edges, square plate;[6] curve 3: stress free edges, square plate;[39] curve 4: stress free edges, infinite plate;[39] curve 5: straight edges, infinite plate; curve 6: stress free edges (eqn (20)), square plate.

For infinite plates with stress-free edges the effects of change in wavelength are not so great. The curve shown for this case was obtained from reference[39] and this analysis had less degrees of freedom than those of Walker or Levy so that the accuracy is not quite so good at loads far beyond buckling.

From the figure it can be concluded that in the case of infinite plates there is very little difference between the effective widths for stiffness obtained for the different edge conditions, and both can be covered adequately by the same curve.

When the half wavelength is allowed to vary then in the case of plates with edges kept straight the effective width for strength is close to that for stiffness and the same curve may be used for both effects. For plates with stress-free edges the maximum stress is relatively independent on half wavelength and eqn (19) is suitable for both square and infinite plates.

Koiter[7] found that for plates with straight edges the effects of rotational restraints at the edges could be taken into account adequately by the same effective width curves so long as the different critical stresses or strains were used. In the case of plates with stress-free edges rotational edge restraints have greater effect. However, if the critical stresses corresponding to different edge restraints are used then the effective width curves for simply supported plates are only slightly conservative for the analysis of rotationally restrained plates and may be applied to these cases with reasonable accuracy.

In the case of plates which have one free edge the growth of very large deflections prohibits the use of very thin plates. Analysis indicates that changes in deflected form occur somewhat more slowly for these plates than for plates supported on both unloaded edges, and that an effective width expression evolved for simply supported plates will be conservative if used for these plates with the appropriate buckling coefficient. In actuality such an effective width expression would be unduly conservative if it were not for some other rather unfortunate behaviour patterns of plates which have a free edge. These will be discussed in greater detail at a later stage.

7. EMPIRICAL EFFECTIVE WIDTH FORMULAE

A large number of effective width formulae have been obtained on the basis of experiment, or a combination of experiment and intuition in the past. In the main these were based on ultimate load conditions so that strength rather than stiffness was the main consideration, and the formulae were often expressed in terms of σ_{CR}, σ_Y and σ_{AV}. A few of these expressions are listed now in terms of effective widths at failure:

Heimerl[40] $\dfrac{b_e}{b} = 0.769 \left(\dfrac{\sigma_{CR}}{\sigma_Y} \right)^{0.2}$ $\rfloor$ and $\lfloor$ sections   (24)

Heimerl[40] $\dfrac{b_e}{b} = 0.794 \left(\dfrac{\sigma_{CR}}{\sigma_Y} \right)^{0.2}$ $-$ $\vdash\!\dashv$ sections   (25)

Schuette[41] $\dfrac{b_e}{b} = 0.8 \left(\dfrac{\sigma_{CR}}{\sigma_Y} \right)^{0.25}$ — $\rbrack$, $\lbrack$ and $\vdash$ sections   (26)

Chilver[14] $\dfrac{b_e}{b} = 0.863 \left(\dfrac{\sigma_{CR}}{\sigma_Y} \right)^{1/3}$ — aluminium channels   (27)

Chilver[14] $\dfrac{b_e}{b} = 0.736 \left(\dfrac{\sigma_{CR}}{\sigma_Y} \right)^{1/3}$ — steel channels   (28)

All of these expressions predict an effective width less than unity when $\sigma_{CR} = \sigma_Y$ thus indicating the effects of inevitable imperfections. It is worthy of note that all the sections involved in the tests were largely composed of plates which had one edge free.

7.1. Effective Widths Based on von Karman's Equation

Equation (1), von Karman's equation, was based on a rather intuitive, if very far sighted, analysis of a simply supported plate. Some points worthy of note are considered below.

This equation implicitly takes variation in buckle half wavelength into account, the variation arising from the analysis being

$$e = \sqrt{\frac{\sigma_{CR}}{\sigma_E}} \qquad (29)$$

At high values of σ_E the equation is very likely to be conservative as no account of bending in the ineffective plate centre was considered in the analysis. However, at low values of σ_E the assumption of a completely ineffective centre does not accurately portray the actual stress distribution and leads to non-conservatism. This may be illustrated by the fact that the tangent effective width at buckling ($\partial \sigma_{AV}/\partial \sigma_{max}$ and E^*/E, since no differentiation between these arises from this equation) is equal to 0·5, whereas rigorous analysis of a plate with no stresses on the unloaded edges gives $E^*/E = 0.408$ and $\partial \sigma_{AV}/\partial \sigma_{max} = 0.26$ at buckling. Of greater significance is the fact that imperfections are not considered, and the combination of these effects leads to unsafe predictions of plate load capacity near the critical load.

On the basis of an extensive experimental investigation, Winter[11] proposed a modification to von Karman's equation which has proved extremely successful in plate design and has formed the basis for many design specifications ever since. Winter's expressions and subsequent suggested modifications are given below. These are expressed here in modified form as functions of σ_{CR}/σ_E with σ_{CR} taken as $4\pi^2 D/b^2 t$ for a plate with both edges supported (stiffened plate) and as $0.425\,\pi^2 D/b^2 t$ for a plate with only one edge supported (unstiffened plate).

$$\frac{b_e}{b} = \sqrt{\frac{\sigma_{CR}}{\sigma_E}}\left(1 - 0.25\,\sqrt{\frac{\sigma_{CR}}{\sigma_E}}\right) \qquad \text{(stiffened plates)} \qquad (30)$$

$$\frac{b_e}{b} = 1.292\,\sqrt{\frac{\sigma_{CR}}{\sigma_E}}\left(1 - 0.326\,\sqrt{\frac{\sigma_{CR}}{\sigma_E}}\right) \qquad \text{(unstiffened plates)} \qquad (31)$$

The coefficient 0·25 obtained from Winter's tests has been subsequently modified to 0·218 in the expression on which the current AISI specification[42] bases effective widths for stiffened plates, and eqn (31) is not now used.

Probably the most significant suggested modification to eqn (30) (or its present equivalent in the AISI Specification) is that due to Faulkner[43] who proposed the form

$$\frac{b_e}{b} = 1.053\,\sqrt{\frac{\sigma_{CR}}{\sigma_E}}\left(1 - 0.263\,\sqrt{\frac{\sigma_{CR}}{\sigma_E}}\right) \qquad (32)$$

The main advantage of this formulation is that this curve has zero slope when $b_e/e = 1$ at $\sigma_E \simeq 0.3\sigma_{CR}$ and thus gives a smooth transition from the $b_e = b$ line which governs the behaviour at stresses lower than this value. The AISI curve, on the other hand, must be replaced by $b_e/b = 1$ when σ_E is less than about $0.46\sigma_{CR}$.

8.   EFFECTIVE WIDTHS USED IN COLD-FORMED STEEL DESIGN SPECIFICATIONS

In the AISI specification the modified version of eqn (30) is used for stiffened plates, i.e.

$$\frac{b_e}{b} = \sqrt{\frac{\sigma_{CR}}{\sigma_E}}\left(1 - 0.218\,\sqrt{\frac{\sigma_{CR}}{\sigma_E}}\right) \qquad (33)$$

with, of course, inbuilt factorisation to ensure safe design.

In the case of unstiffened plates this formulation is no longer used and more severe limitations on design loads than would be obtained using such a formulation are now applied.

In the Swedish specification, as reviewed in references 44 and 45 eqn (33) is used to determine the effective width for strength of stiffened plates.

For evaluation of deflections the effective width for stiffness is taken as

$$\frac{\bar{b}_e}{b} = 0 \cdot 827 \left(\frac{\varepsilon_{CR}}{\varepsilon_x} \right)^{1/3} \tag{34}$$

In this specification a path whereby $\bar{b}_e/b$ and b_e/b become equal at the ultimate load is set up, although it is rather difficult to see any theoretical justification for such a situation.

In the case of unstiffened plates the relevant expressions in terms of σ_{CR} and σ_E are

$$\frac{b_e}{b} = \sqrt{\frac{\sigma_{CR}}{\sigma_E}} \tag{35}$$

$$\frac{\bar{b}_e}{b} = 1 \cdot 25 \sqrt{\frac{\sigma_{CR}}{\sigma_E}} \tag{36}$$

In the above expressions provision is made for variation of the buckling coefficient depending on rotational restraints, so that these equations are valid for any degree of rotational restraints at the plate edges.

The equations relating to stiffened plate behaviour have also been incorporated in European recommendations for the design of profiled sheeting.[46]

In the specification of the Canadian Standards Association the effective width expression used is a simple modification of von Karman's original expression, i.e.

$$\frac{b_e}{b} = 0 \cdot 863 \sqrt{\frac{\sigma_{CR}}{\sigma_E}} \tag{37}$$

The UK specifications for cold-formed steel sections gives stress reduction factors rather than effective width factors, but these are interchangeable with effective width factors, and are used as effective width factors in evaluating beam deflections. Prior to 1975 the relevant equation used was a conservative modification of the experimentally derived equations of Chilver, eqns (27) and (28), i.e.

$$\textit{stress reduction factor, } CL \left(\equiv \frac{b_e}{b} \right) = 0 \cdot 66 \left(\frac{\sigma_{CR}}{\sigma_Y} \right)^{1/3} \tag{38}$$

In the present specification[23] the values of CL, or b_e/b, are based on the plate analysis of Walker.[22] This analysis incorporates an imperfection amplitude which is geared to accurately describe the variation of typical imperfections obtained in sections with different material width to thickness ratios and yield stresses.

On this basis the factors used in this specification may be claimed to give the most accurate prediction of single plate behaviour. There are drawbacks however in that the governing equations are substantially more complex than the effective width equations of other specifications, and the factors are presented in tabular form rather than as an equation. This can cause some difficulty in, for example, routine computer aided design. Since in general restraints between adjacent plates of cold-formed sections tend to increase the effective widths in practice there is a case for the adoption of a relatively simple expression of less theoretical exactitude.

The variation of effective widths for strength given by the various codes are illustrated in Fig. 14. The equation of von Karman and Faulkner's modification of Winter's expression are also included for comparison.

With the exception of the curve of reference,[23] for which no simple expression is available, none of the expressions apply to very low values of stress and must be replaced by the expression $b_e/b = 1$ at some limiting value of σ_E/σ_{CR}. It would appear that the UK specification more realistically describes the effective widths at low stresses than the other formulations. At higher stresses, however, this effective width curve has

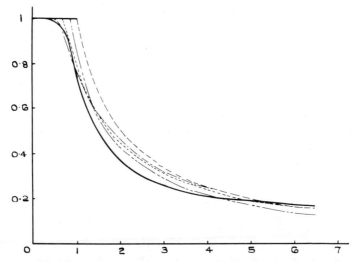

FIG. 14.   Effective width curves used in design. Ordinate: b_e/b. Abscissa: σ_Y/σ_{CR}. (--------), AISI; (- · · · · · · -), CSA; (———), UK; (– · – · –), Faulkner; (– – – –), von Karman.

been shown to consistently describe the lower bound values of experimental results whereas the other curves tend to describe the average values obtained from tests.

9. SOME SUGGESTED EXPRESSIONS FOR DESIGN USE

The effective width curves based on Winter's initial expression have been well-proven in accuracy and conservatism over a large range of plate width to thickness ratios. It is in the region of low strains in which these curves have their least satisfactory shape, and could be improved upon, and an expression which is valid for all values of stress would have some advantages.

A suitable form of expression, which is applicable for all values of σ_{max} is

$$\frac{b_e}{b} = \left(1 + C\left(\frac{\sigma_{max}}{\sigma_{CR}}\right)^m\right)^{-n} \tag{39}$$

Using such an expression with suitable values of C, m and n furnishes curves which have $b_e/b = 1$ at $\sigma_{max}/\sigma_{CR} = 0$, and can be easily fitted to existing curves. Suitable values of these coefficients are suggested for different cases.

9.1. Effective Width for Strength—Stiffened Plates
In this case suitable values are $C = 4$, $m = 3$, $n = 0.15$, giving the equation

$$\frac{b_e}{b} = \left(1 + 4\left(\frac{\sigma_{max}}{\sigma_{CR}}\right)^3\right)^{-0.15} \tag{40}$$

This curve is shown in Fig. 15 together with those of the AISI and UK specifications. At high stresses there is good agreement with the Winter type curves and at low stresses good agreement with that of reference 23. In the inset to this figure the effective widths at relatively low stresses are compared with proposed design curves for bridge plating[47,48] and the form and magnitudes are in good agreement. This accuracy at low stresses allied to the applicability of the equation at all strains is a point of advantage over existing expressions.

9.2. Effective Width for Stiffness—Stiffened Plates
The majority of the empirical effective width expressions are for strength determination, and although specifications make use of these for stiffness

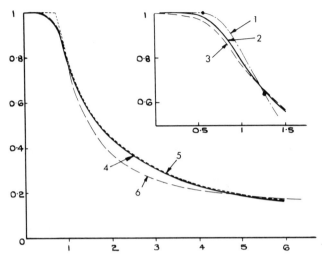

FIG. 15. Comparison of curve from eqn (40) with existing curves. For both graphs: ordinate, b_e/b; abscissa, $\sqrt{\sigma_Y/\sigma_{CR}}$. (The abscissa of the smaller graph has an expanded scale.) Curve 1, from reference 48; curve 2, from eqn (40); curve 3, from reference 47; curve 4, from eqn (40); curve 5, according to AISI;[42] curve 6, according to UK specifications.[23]

evaluation also it has been found on a number of occasions that undue conservatism, or overestimation of deflections, often results (see, for example, reference 44).

As has been discussed already there is strong evidence that for displacement purposes the effective width varies as the cube root of $\varepsilon_{CR}/\varepsilon_x$. This has been given experimental substantiation by the results of Thomasson[44] who found that good agreement with experimental behaviour was given by an equation very close to eqn (34).

If the index -0.15 is changed to $-1/9$ in eqn (40), then at high stresses the effective width will vary as the cube root of σ_{CR}/σ_{max}. Although it is more correct to state effective widths for stiffness in terms of strains rather than stresses, it is more convenient to use a function of very similar form to that used for stresses. Also since σ_{max}/σ_{CR} is always greater than $\varepsilon_x/\varepsilon_{CR}$ an effective width of the form of eqn (40) with modified index will give slightly more conservative results than if σ_{max}/σ_{CR} is replaced by $\varepsilon_x/\varepsilon_{CR}$. The suggested effective width for displacement evaluation is, therefore

$$\frac{\bar{b}_e}{b} = \left(1 + 8\left(\frac{\sigma_{max}}{\sigma_{CR}}\right)^3\right)^{-1/9} \tag{41}$$

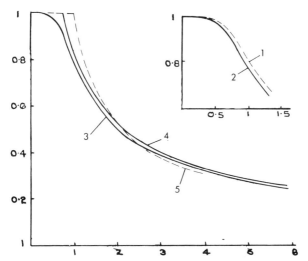

Fig. 16. Comparison of curve from eqn (41) with existing curves. For both graphs: ordinate, $\bar{b}_e/b$; abscissa, σ_m/σ_{CR}. (The abscissa of the smaller graph has an expanded scale.) Curve 1, from reference 47: curve 2, from eqn (41); curve 3, from eqn (41); curve 4, from eqn (34); curve 5, from reference 39, Fig. 13.

Figure 16 shows this effective width curve compared with that of Thomasson, eqn (34) and of the lowest curve of Fig. 13 for infinite plates. The inset shows comparison at low stresses with the stiffness curve suggested for box girder bridge design in reference 47. As can be seen, agreement is good with Thomasson's curve, and that of Fig. 13 at high stresses, and at low stresses there is slight conservatism in comparison with that of reference 47.

9.3. Unstiffened Plates

In the case of plates which have one free edge, i.e. 'outstands' or 'unstiffened plates', there are at present wide variations in different design specifications. As previously mentioned the AISI specification, which in earlier editions had treated these in a similar way to stiffened plates, now permits only very limited post-buckling strength to be used for part of the range of width to thickness ratios covered, and in fact does not allow the theoretical buckling load to be reached for most of the range considered. The UK specification on the other hand, gives factors based on exactly the same expressions as for stiffened plates, with the buckling coefficient changed to suit, and the Swedish specification actually allows higher effective widths for a given ratio of σ_E/σ_{CR} than for stiffened plates.

There are special problems caused by the existence of the free edge for this type of element, and this will be discussed in the next section. The effective widths postulated for this case in this section apply only in the situation where uniform compression along the element can be maintained. Under these circumstances it has been observed earlier that at least in the initial post-buckling range unstiffened plates have higher effective widths for stiffness and strength than stiffened plates. It is also worth noting that the empirical effective widths quoted in eqns (24)–(28) were largely based on sections substantially composed of unstiffened elements, and these also give relatively high values of effective width at stresses far beyond buckling.

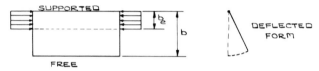

FIG. 17.   Unstiffened plate.

A simple estimate of the behaviour at loads far beyond buckling can be obtained using a similar type of analysis as that used for simply supported plates by von Karman *et al.*, from a consideration of a simply supported free plate which is effective in withstanding load only near the supported edge as shown in Fig. 17. Now assuming that the deflections across the plate vary linearly and performing a simple buckling analysis produces the result

$$\sigma_E = \sigma_{CR}\left(\frac{b}{b_e}\right)^3 \tag{42}$$

where σ_{CR} is the buckling stress under uniform load across the plate. This gives

$$\frac{b_e}{b} = \left(\frac{\sigma_{CR}}{\sigma_E}\right)^{1/3} \tag{43}$$

Because of the analysis used this equation will be conservative at high stresses, although the effective widths given are higher than those of von Karman for stiffened plates. At stresses near buckling the inaccuracies of von Karman's derivation, due to the unrealistic stress distribution and neglect of imperfections, are repeated here.

Observing the values of eqns (24)–(28) suggests that the effective width may be approximated to by the equation

$$\frac{b_e}{b} = \left(1 + 8\left(\frac{\sigma_{max}}{\sigma_{CR}}\right)^3\right)^{-1/9} \tag{44}$$

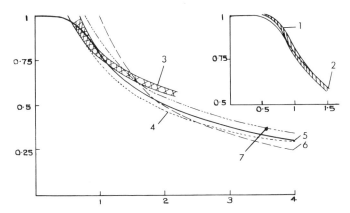

FIG. 18. Comparison of curve from eqn (44) with existing curves. For both graphs: ordinate, b_e/b; abscissa, $\sqrt{\sigma_Y/\sigma_{CR}}$. (The abscissa of the smaller graph has an expanded scale.) Curve 1, from reference 32; curve 2, from eqn (44); curve 3, from eqns (24)–(26); curve 4, from eqn (28); curve 5, from eqn (44); curve 6, from eqn (35); curve 7, from eqn (27).

i.e. the same as the stiffness effective width for stiffened plates. The effective widths obtained from eqn (44) are compared with the empirical eqns (24)–(28) and with those of the Swedish code in Fig. 18, indicating that this equation prescribes something of an average value of those from other sources. In the inset comparison is made with proposed curves of reference 32 for outstands in stiffened plating, giving good agreement.

On the basis of the findings earlier in this chapter, eqn (44) would also appear to be quite accurate in the assessment of b_e/b and $\bar{b}_e/b$ for simply supported plates with the unloaded edges constrained to remain straight.

10. SPECIAL PROBLEMS OF UNSTIFFENED PLATES

While this type of plate is more efficient after buckling than a stiffened plate if uniform compression can be maintained, the highly eccentric shedding of stresses after buckling can cause problems. A rigorous analysis of this type of problem is given in reference 25, but for the purposes of illustration the results of a simpler analysis, which is still of good accuracy, are shown here.

For the plate loaded by moment M and axial force P as shown in Fig. 19(a), if it is assumed that stress variation along the plate and shear stresses are negligible and the deflections vary linearly across the plate, the

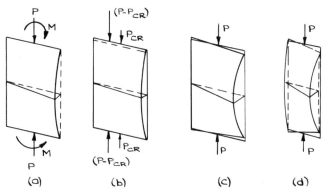

FIG. 19.   Representation of unstiffened plate behaviour.

following equations can be derived, neglecting imperfections, for the post-buckling range

$$\frac{P}{P_{CR}} = \frac{4}{9}\frac{u}{u_{CR}} - \frac{13}{16}\frac{\theta b}{u_{CR}} + \frac{5}{9} \tag{45}$$

$$\frac{M}{bP_{CR}} = -\frac{5}{36}\frac{u}{u_{CR}} + \frac{17}{144}\frac{\theta b}{u_{CR}} + \frac{5}{36} \tag{46}$$

Before buckling the corresponding equations are

$$\frac{P}{P_{CR}} = \frac{u}{u_{CR}} + \frac{\theta b}{2u_{CR}} \tag{47}$$

$$\frac{M}{bP_{CR}} = \frac{1}{12}\frac{\theta b}{u_{CR}} \tag{48}$$

in which P_{CR} and u_{CR} are the values of buckling load and end displacement at buckling obtained for a uniformly compressed plate, and θ is the angle through which the plate is bent in its own plane.

Examination and manipulation of these equations gives the following results:

Under uniform compression, $\theta = 0$, the tangent effective width after buckling is $4/9b$, the same value as that obtained by Stowell. To ensure uniform compression a moment is required which causes the resultant load to move towards a position $3/16b$ from the supported edge

Fig. 19(b). If no moment is applied ($M = 0$) then severe in plane bending is induced due to the eccentricity of resistance offered by the plate, and the displacement of the free edge occurs at a rate of 51 times its pre-buckling rate, while that of the load point increases at a rate of 21 times its prebuckling rate and displacements, strains, stresses, etc., at the supported edge actually reduce and become tensile after buckling (Fig. 19(c)). Therefore under central load only, without either forced or natural internal repositioning of the load point the plate is highly ineffective after buckling and post-buckling resistance is negligible.

Under pure moment, or combinations of M and P, the neutral axis of bending after buckling is $3/16b$ from the supported edge, and the flexural rigidity of the plate in its own plane, EI, reduces to $1/16$ its former value. This raises the possibility of Euler buckling of such a plate in its own plane (Fig. 19(d)).

The Euler stress for such a plate, σ_{EU}, may be related to the local buckling stress by the expression

$$\frac{\sigma_{EU}}{\sigma_{CR}} = 2 \cdot 14 \left(\frac{b}{L}\right)^2 \left(\frac{b}{t}\right) \tag{49}$$

If the plate has buckled the reduced Euler load, σ_{RE} corresponding to the reduced flexural stiffness is $1/16$ of this value. Therefore for a plate length L without any support from adjacent elements of a section, it is found that if L/b is greater than $1 \cdot 464b/t$ then σ_{EU} is less than σ_{CR}. If $L/b >$ $0 \cdot 366b/t$ then $\sigma_{RE} < \sigma_{CR}$. For values of L/b between these limits local buckling will be unstable and possibly dynamic with rapid failure immediately at buckling. For lower values of L/b some post-buckling strength will be available, but for higher values then the unstiffened element is completely dependent on support from adjacent elements.

The approach used here is perhaps simplistic in its neglect of imperfections, assistance from adjacent elements, etc. Nevertheless it illustrates the problems which can arise with unstiffened elements. The effects of this type of behaviour are illustrated in references 49 and 50 for cases of bridge plating and plain channel sections, respectively. While in some cases use can be made of the high post-buckling strength and stiffness of these elements subject to uniform compression, for example with symmetrically disposed unstiffened elements, or with these elements in line with stiffened elements, for safe design in the general case the severe approach of the AISI specification is to be recommended.

11. PLATES UNDER COMBINED BENDING AND COMPRESSION

In recent years the use of wide thin webs has led to the requirement in some cases to design these webs for use in the post-buckling range, for example in the design of profiled sheeting.

The success of effective width formulations in dealing with uniformly compressed plates suggests that this method of approach may be used under bending, or combinations of bending and compression, and a number of investigators have examined the applicability of effective width type analysis under these conditions. The results of some of these investigations are given in references 44 and 45 and in the latter paper it is shown that this can produce good agreement with experimental failure loads for profiled sheeting.

While available space does not permit a detailed discussion of this topic the writer must confess to having some reservations about the use of effective widths in this context. Because of the varying stress across the section the widths and position of the effective parts of the plate must be decided, and these depend on the depth of the compression zone which is itself dependent on the effective width. As a result the analysis becomes somewhat complex and the simplicity of the effective width approach is negated. Furthermore, while the effective width concept can exactly model the strength and stiffness behaviour of a uniformly compressed plate, the same is not true where bending is concerned, and the plate behaviour cannot be adequately modelled using this concept. Thomasson[44] refers to the work of Kloppel and Bilstein who make this claim, and this is also the experience of the writer in an investigation under progress.

In the case of the pure bending of simply supported plates, for example, analysis shows that for a perfect plate the instantaneous or tangent flexural stiffness after buckling is 0·71 times that before buckling, and in-plane bending after buckling occurs about an instantaneous neutral axis position $0·358b$ from the tension edge. If the effective width approach is applicable after buckling it must also be applicable to the instantaneous effects, and it is found that no combination of effective widths situated at the compression and tension edges can produce the given neutral axis position and flexural stiffness simultaneously. The rate of growth of maximum compressive stress for this case increases at buckling to 2·5 times its pre-buckling value.

Postulation of flexural stiffness and neutral axis position, together with compressional stiffness in the case of bending and axial load, would seem to be a more direct approach to this problem, and lead to simpler analysis. Equations similar to those illustrated, given for pure compression earlier in the chapter, can be obtained, and imperfection effects can be taken into account in the same way as for pure compression.

12. APPLICATION OF EFFECTIVE WIDTH EXPRESSIONS IN THE PLASTIC RANGE

The behaviour of plates in the elasto-plastic range have become increasingly important in recent years, particularly in the case of plates which have buckling and yield stresses of the same order of magnitude, i.e. in the approximate range $\frac{1}{4}\sigma_Y < \sigma_{CR} < 4\sigma_Y$. The behaviour in this range is important in heavy civil engineering structures where the weight saving advantages of thin plating are now being exploited more than in the past, and in the cold-formed sections field where methods of utilising the post yield capacity of beams with relatively thick elements are of interest.

It has been found (see, for example, references 18 and 34) that elastically derived effective widths can be used to give predictions of good accuracy in the plastic range if used in conjunction with the material stress–strain curve. This has been examined recently in reference 52 where it is found that analysis in the elasto-plastic range using elastically derived effective widths obtained using equations similar to eqns (11)–(13) produced excellent

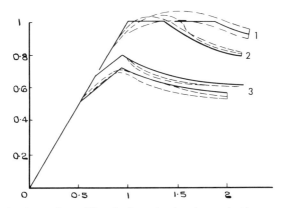

FIG. 20. Behaviour of unstiffened plates in the elasto-plastic range. Ordinate: σ_{AV}/σ_Y. Abscissa: $\varepsilon/\varepsilon_Y$. Curves 1, $b/t = 13\cdot3$; curves 2, $b/t = 15\cdot7$; curves 3, $b/t = 21\cdot9$; curve 4, $b/t = 24\cdot2$. (———), simply supported free plates, $a/b = 8$, effective width analysis; (----), experiment results (see reference 32).

agreement, both with rigorous elasto-plastic analysis and experiment, throughout the complete range investigated. An example of the correlation with experiment is shown in Fig. 20 where the predictions based on elastic effective widths for simply supported free plates are compared with experiments by Rogers and Dwight.[32] Very good agreement is found for all plates.

This type of analysis is applied to beam compression elements in reference 52 and indicates the increases in calculated load capacity which can be obtained using the inelastic reserve of strength.

It would on the whole be expected that elastic effective widths apply in the plastic range, despite the absence of mathematical justification, for many cases of plates with simply supported edge conditions. For plates with fixed or rotationally restrained edges additional complications arise because the presence of plasticity reduces the restraining moments on the edges substantially. If the plates are very thin, so that σ_{CR} is significantly less than σ_Y, failure of plates with stress-free edges occurs by a different mechanism than for thicker plates and it would not be expected that elastic effective widths could adequately describe post yield behaviour in this case.

13. EFFECTIVE WIDTHS FOR SPECIALLY ORTHOTROPIC PLATES

The behaviour after buckling, of glass fibre plates and various other laminated composite plates, has received substantial attention over the past few years. As these plates in general have different flexural and axial stiffnesses in different directions they must be dealt with using anisotropic plate theory and the equations given here do not apply.

An important case is that of specially orthotropic plates in which the principal stiffnesses are aligned in the directions of the plate edges. For multi-layer plates of this type the ratio of flexural stiffnesses in the principal directions is the same as the ratio of axial stiffnesses in these directions. In a recent paper[38] it has been shown that for plates of this type there is a basis for using effective widths derived for isotropic plates in design. While the behaviour of a highly orthotropic plate can be very much different from that of an isotropic plate if they both have the same buckle half wavelength, there is close correlation of the post-buckling behaviour evaluated on the basis that the natural half wavelength occurs in all cases. The natural half wavelength, i.e. the wavelength which a very long plate would naturally assume, is highly dependent on the degree of orthotropy. If the effective widths are evaluated for infinitely long plates then it is found that the effective widths evaluated for isotropic plates are applicable for stiffness evaluation of specially orthotropic plates throughout the post-buckling range.

Because of the highly complex nature, and to an extent process dependency, of failure in composite plates it is unlikely that an effective width formulation could adequately predict the strength of composite plates with any degree of generality. An indication of some of the failure characteristics is given in reference 53.

14. CONCLUDING REMARKS

In this chapter an attempt has been made to describe the effective width analysis of plates, its historical background and its application to various problems. There has been so much written about this topic in the past that only a small fraction of the published literature is included in the references. Some aspects have been treated only in a rather cursory fashion while some have been omitted completely.

The form of the design effective widths suggested has been based only on general accuracy, and not on ease of applicability, although the expressions are not complicated. For many problems the use of instantaneous or tangent values of effective widths or section properties can simplify analysis considerably (see, for example, reference 50), and in such cases the type of expression obtained for the initial post-buckling range, which uses fixed values of tangent stiffness, etc., is ideally suited. It is sometimes even useful to approximate other effective width expressions by fitting expressions of this type over part of the range to allow the simple analysis procedures to be used.

The range of width to thickness ratios for which the various design expressions are applicable is up to 500 for the AISI and similar expressions for stiffened steel plates, and this has been thoroughly verified by experiment. There is also evidence to suggest that such expressions can give conservative predictions outwith this range, so long as the large deflections can be lived with. However it is also known that other effects which are secondary and neglected, within the range, sooner or later become predominant, and the basic equations become inadequate. In view of this any extension of the range of validity would require thorough experimental substantiation.

In the case of steel plates the effective widths at failure obtained by any of the design formulae are in the region of 40–55 times the plate thickness if σ_Y is greater than twice the critical stress and generally close to the figure of $50t$ mentioned at the beginning of this chapter.

REFERENCES

1. SCHUMAN, L. and BACK, G., Strength of rectangular flat plates under edge compression, *NACA Rep. No.356*, 1930.
2. VON KARMAN, T., SECHLER, E. E. and DONNEL, L. H., Strength of thin plates in compression, *Trans. ASME*, **54**, 1932.
3. COX, H. L., Buckling of thin plates in compression, *ARC R & M No.1554*, 1934.
4. COX, H. L., The theory of flat panels buckled in compression, *ARC R & M No.2178*, 1945.
5. MARGUERRE, K., The apparent width of the plate in compression, *NACA TM No.833*, 1937.
6. LEVY, S., Bending of rectangular plates with large deflections, *NACA Rep. No.737*, 1942.
7. KOITER, W. T., The effective width of infinitely long flat rectangular plates under various conditions of edge restraint, *NLL Report S287, Nat. Aero. Res. Inst.*, Amsterdam, 1943.
8. HEMP, W. S., The buckling of a flat rectangular plate in compression and its behaviour after buckling, *ARC R & M No.2041*, 1945.
9. VAN DER NEUT, I. A., in: *Proc. 7th Int. Congress for Appl. Mech.*, Vol. 1, 1948.
10. HU, P. C., LUNDQUIST, E. E. and BATDORF, S. B., Effect of small deviations from flatness on effective width and buckling of plates in compression, *NACA TM No.1124*, 1946.
11. WINTER, G., Strength of thin steel compression flanges, *Cornell Univ. Eng. Exp. Stn, Reprint No.32*, 1947.
12. KENEDI, R. M., SHEARER SMITH, W. and FAHMY, F. O., *Trans. Inst. of Engrs and Shipbuilders of Scotland*, **99**(4), 1955.
13. CHILVER, A. H., The stability and strength of thin-walled steel struts, *The Engineer*, 1953.
14. CHILVER, A. H., *Civil Eng. and Public Works Review*, **48**(576), 1953.
15. HARVEY, J. M., Structural strength of thin-walled channel sections, *Engineering*, **CLXXV**, 1953.
16. STOWELL, E. Z., Compressive strength of flanges, *NACA Tech. Rep. No.1029*, 1951.
17. COAN, J. M., Large deflection theory for plates with small initial curvature loaded in edge compression, *Trans. ASME*, **73**, 1951.
18. BOTMAN, M. and BESSELING, J. F., The effective width in the plastic range of flat plates under compression, *NLL Rept. 5445*, Amsterdam, 1954.
19. STEIN, M., Loads and deformations in buckled rectangular plates, *NASA Tech. Rep. R-40*, 1959.
20. YAMAKI, N., The post-buckling behaviour of rectangular plates with small initial curvature loaded in edge compression, *J. of App. Mech.* **26**, 1959.

21. GRAVES SMITH, T. R., in: *Thin-Walled Steel Structures*, C. K. Rockey and H. V. Hill, Eds., Crosby Lockwood and Son Ltd, London, 1967.
22. WALKER, A. C., *Aero Quarterly*, **XX**, August 1969.
23. BRITISH STANDARDS INSTITUTION, *Specification for the use of cold-formed steel sections in building*, BS 449, Addendum No. 1, 1975.
24. RHODES, J. and HARVEY, J. M., Plates in uniaxial compression with various support conditions at the unloaded boundaries, *Int. J. Mech. Sci.*, **13**, 1971.
25. RHODES, J., HARVEY, J. M. and FOK, W. C., The load carrying capacity of initially imperfect eccentrically loaded plates, *Int. J. Mech. Sci.*, **17**, 1975.
26. RHODES, J. and HARVEY, J. M., *J. of Eng. Mech. Div. ASCE*, **103**(EM3), June 1977.
27. WILLIAMS, D. G. and WALKER, A. C., *Proc. Inst. Civ. Engrs*, **59**(2), December 1975.
28. MOXHAM, K. E., Theoretical determination of the strength of welded steel plates under in-plane compression, *Cambridge Univ., Rep. CU ED/C-Struct/TR65*, 1971.
29. FRIEZE, P. A., DOWLING, P. J. and HOBBS, R. E., in: *Steel Plated Structures*, P. J. Dowling, J. E. Harding and P. A. Frieze, Eds., Crosby Lockwood Staples, London, 1977.
30. CRISFIELD, M. A., Ivanov's yield criterion for thin plates and shells using finite element, *Transport and Road Research Laboratory, Rep. LR919*, Crowthorne, 1979.
31. LITTLE, G. H., Rapid analysis of plate collapse by live energy minimisation, *Int. J. Mech. Sci.*, **19**, 1977.
32. ROGERS, N. A. and DWIGHT, J. B., in : *Steel Plated Structures*, P. J. Dowling, J. E. Harding and P. A. Frieze, Eds., Crosby Lockwood Staples, London, 1977.
33. JOMBOCK, J. P. and CLARK, J. W., *J. of Struct. Div. ASCE*, **87**(ST5), 1967.
34. KOITER, W. T., Introduction to the post-buckling behaviour of flat plates, *Mem. Soc. R. Sci.*, Vol. 8, Liege, 1963.
35. SUPPLE, W. J. and CHILVER, A. H., in: *Thin-Walled Structures*, A. H. Chilver, Ed., Chatto and Windus, London, 1967.
36. BULSON, P. S., *The Stability of Flat Plates*, Chatto and Windus, London, 1970.
37. SUPPLE, W. J., Changes of waveform of plates in the post-buckling range, *Int. J. Solids and Structures*, **6**, 1970.
38. RHODES, J. and MARSHALL, I. H., in: *Composite Structures* (Proc. Int. Conf. on Composite Structures, Paisley, 1981), I. H. Marshall, Ed., Applied Science Publishers, London, 1981.
39. RHODES, J. and HARVEY, J. M., *J. Mech. Eng. Sci.*, **13**(2), 1971.
40. HEIMERL, G. J., Determination of plate compressive strength, *NACA Tech. Note No. 1480*, 1947.
41. SCHUETTE, E. H., Observations on the maximum average stress of flat plates buckled in edge compression, *NACA Tech. Note No. 1625*, 1947.
42. AMERICAN IRON AND STEEL INSTITUTE, *Specification For The Design of Cold-Formed Steel Structural Members*, AISI, New York, September 1980.
43. FAULKNER, D., Discussion on 'Ultimate longitudinal strength' by J. B. Caldwell, *Trans. Royal Inst. of Naval Architecture*, **107**, 1965.
44. THOMASSON, P. O., Thin-walled C-shaped panels in axial compression, *Swedish Council for Building Research, Document D1*, 1978.

45. BERGFELT, A. and EDLUND, B., in: *Thin-Walled Structures*, J. Rhodes and A. C. Walker, Eds., Granada, London, 1980.
46. ECCS, Provisional European recommendations for the design of profiled sheeting and sections, part 1—profiled sheeting, *ECCS Comm. 17, Working Group 1*, 1979.
47. CHATTERJEE, S. and DOWLING, P. J., in: *Steel Plated Structures*, P. J. Dowling, J. E. Harding and P. A. Frieze, Eds., Crosby Lockwood Staples, London, 1977.
48. SCHEER, J. and NOLKE, H., in: *Steel Plated Structures*, P. J. Dowling, J. E. Harding and P. A. Frieze, Eds., Crosby Lockwood Staples, London, 1977.
49. FOK, W. C., RHODES, J. and WALKER, A. C., *Aero Quarterly*, **XXVII**, May 1976.
50. RHODES, J. and HARVEY, J. M., *2nd Int. Colloquium on Stability of Steel Structures, Preliminary Report*, April 1977.
51. RHODES, J., *Proc. Inst. Civ. Engrs.*, **71**(2), March 1981.
52. RHODES, J. and MARSHALL, I. H., *Proc. 5th Int. Speciality Conference on Cold-Formed Steel Structures*, University of Missouri-Rolla, November 1980.
53. BANKS, W. M., *PhD Thesis*, University of Strathclyde, 1977.

Chapter 5

CONNECTIONS IN THIN-WALLED STRUCTURES

J. W. B. STARK and A. W. TOMÀ

Department of Steel Structures,
Institute TNO for Building Materials and Building Structures,
Delft, The Netherlands

SUMMARY

In this chapter attention is given to the structural requirements of connections. Although the non-structural requirements are equally important, these are not discussed in detail. A survey is given of the most frequently used fastener types in thin-walled structures. Thereafter, the principles of design, the forces in the connections and the strength of fastenings are considered. As a tool for designers, formulae for determining the design strength of fastenings are given in Section 5.4.

1. INTRODUCTION

Connections are an important part of every structure not only from the point of view of structural behaviour, but also in relation to the cost of production. It has been shown that for a structure of hot-rolled sections about 40 % of the total costs are directly or indirectly influenced by the connections. There is no reason to believe that this part will be much smaller for light-weight structures. For economic reasons the influence of the joining process on the costs will tend to increase.

A variety of joining methods are available for light-weight structures. Correct selection is governed by a large number of factors (see Scheme 1). This chapter focuses attention on the structural requirements, but this does

160   J. W. B. STARK AND A. W. TOMÀ

Structural Requirements
1. Strength
2. Stiffness
3. Deformation capacity

Non-structural Requirements
1. Economic aspects, such as:
   (a) total number of fastenings which have to be made;
   (b) skill required;
   (c) ability to be dismantled;
   (d) design life;
   (e) installed costs of the fastening. The cost factors are:
       (i)   fastener piece part cost;
       (ii) direct labour cost;
       (iii) indirect labour cost;
       (iv) application tools cost;
       (v)   maintenance cost;
       (vi) inventory cost.
2. Durability, which depends on:
   (a) chemical aggressiveness of the environment;
   (b) possible galvanic corrosion;
   (c) stress corrosion (can be important with elevated temperatures and aggressive chemical environments).
3. Watertightness
4. Aesthetics

SCHEME 1.   Requirements for connections in thin-walled structures.

not mean that in the opinion of the authors the structural behaviour is the only important factor.

For a long time steel structures were composed of hot-rolled steel members. These members had relatively high plate thicknesses. The cold-formed structural members differ from hot-rolled structural members by their reduced thicknesses, the forming process and the form of the sections. Therefore, the structural properties of hot-rolled and cold-formed members differ too. Mechanical properties of the material change during cold-forming. The structural properties of members in bending or compression depend on material thickness and the form of the section.

Fasteners used for connections of hot-rolled sections are especially developed to be used in thicker plates. These fasteners can also be used for thinner plates, however, sometimes with a different structural behaviour. Furthermore special fasteners are available which are sometimes preferable because of simplicity of assembly or for economic reasons.

2. TYPES OF CONNECTIONS

In building skins made of profiled steel sheeting such as floors, roofs and walls a fairly large number of connections will be necessary (Fig. 1), i.e.:

—connection of the sheet and the supporting member;
—fastening in the seam of two sheets;
—fasteners to fix the insulating material to the steel sheet.

The connection is an important link in the chain of transfer of forces from the sheet to the main steel structure. Frequently, the steel sheeting is used as

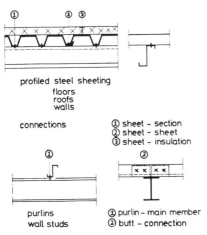

profiled steel sheeting
floors
roofs
walls

connections

① sheet – section
② sheet – sheet
③ sheet – insulation

purlins
wall studs

② purlin – main member
② butt – connection

FIG. 1.   Connections in floors, roofs and walls. (Reproduced from *Sheet Metal Industries*, **54**(2), 108–20, 1977; with permission, Fuel and Metallurgical Journals Ltd.)

lateral bracing for the purlins and sometimes as the bracing component for the stability of the whole structures (stressed skin design). These applications have influence on the requirements for the fastenings. Cold-formed steel sections are used in large quantities for purlins and wall studs. Relevant connections in these applications are:

—the connection between the purlin and the main member;
—the butt-joints in the purlins.

Figure 2 shows other kinds of connections in built-up members to connect two cold-formed sections together or to strengthen a cold-formed

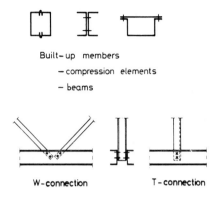

FIG. 2.   Connections in thin-walled sections. (Reproduced from *Sheet Metal Industries*, 54(2), 108–20, 1977; with permission, Fuel and Metallurgical Journals Ltd.)

section with a flat sheet. These members are used for compression elements and for beams. Some types of connections as used in trusses and frames are shown schematically below, although the list is not complete.

3.   TYPES OF FASTENERS OR FASTENING METHODS

A review of the fastening methods available is shown in Scheme 2. The most appropriate application is given for each fastener. In this context thin material is less than 2 mm and thick material more than 2 mm.

3.1.   Welding
Two types of welding can be used for connections in thin-walled structures: resistance (line or spot) and fusion. Both types can be used for connecting thin to thin material and for thin to thick material.

3.2.   Bolts
The most usual application of bolts is in sections; so they are used for connecting thick to thick material. A friction grip fastening in galvanised material with a thickness less than 8 mm is not recommended—A loss of the pre-stressing force will appear, caused by creep of the zinc layer (see reference 14).

Welding $\left\{\begin{array}{l}\text{resistance welding}\\\text{fusion welding}\end{array}\right.$

Bolts $\left\{\begin{array}{l}\text{black bolts}\\\text{friction grip bolts}\\\text{huckbolts}\\\text{special types}\end{array}\right.$

Screws $\left\{\begin{array}{l}\text{thread forming}\\\text{thread cutting}\\\text{self-drilling–self-tapping}\end{array}\right.$

Blind rivets $\left\{\begin{array}{l}\text{open}\\\text{closed}\end{array}\right.$

Powder actuated fasteners (nails)

Special types $\left\{\begin{array}{l}\text{seaming}\\\text{brazing}\end{array}\right.$

SCHEME 2.   Review of types of fasteners or fastening methods.

Figure 3 shows two Huckbolt types and Fig. 4 shows three types of special nuts.

3.3. Screws
The most usual application of screws is for fastening thin to thin material and thin to thick material.

Figure 5 shows three types of screws with different threads. Each type is

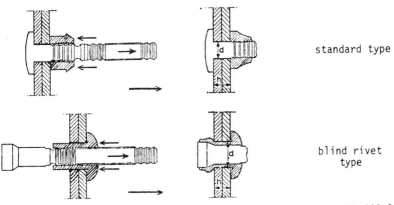

standard type

blind rivet type

FIG. 3.   Huck bolts. (Reproduced from *Sheet Metal Industries*, **54**(2), 108–20, 1977; with permission, Fuel and Metallurgical Journals Ltd.)

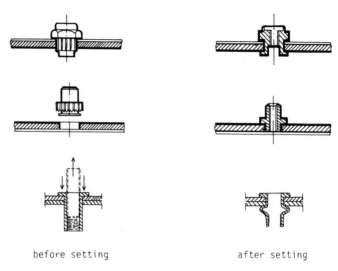

before setting   after setting

FIG. 4.   Special nuts.

available with various forms of head (hexagon head, hexagon washer head, pan head with a slot or driving-recess, countersunk head with a slot or driving-recess).

3.4. Blind Rivets

The most usual application of blind rivets is for fastenings between thin materials. Thin to thick material connections are also possible in certain circumstances (when deformation capacity is not required). Blind rivets are available in steel, aluminium, monel and corrosion resistant steel. Figure 6 shows two types of blind rivets, open and closed.

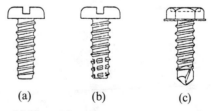

(a)   (b)   (c)

FIG. 5.   Screw types. (a) Thread forming (type B); (b) thread cutting (type BF); (c) self-drilling. (Reproduced from *Sheet Metal Industries*, **54**(2), 108–20, 1977; with permission, Fuel and Metallurgical Journals Ltd.)

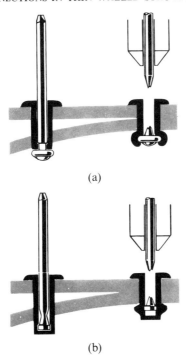

(a)

(b)

FIG. 6. Two blind rivet types. (a) Open blind rivet; (b) closed blind rivet. (Reproduced from *Sheet Metal Industries*, **54**(2), 108–20, 1977; with permission, Fuel and Metallurgical Journals Ltd.)

3.5. Powder Actuated Fasteners

The most usual application of powder actuated fasteners is for the fastening of thin to thick material (e.g. fastening of profiled steel sheeting). This type normally requires a minimum thickness of 6 mm of the supporting steel. Figure 7 shows an assembled fastener. (Also a nail with a sealing cap is shown. The sealing cap will be placed after driving the nail.)

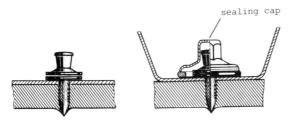

FIG. 7. A powder actuated fastener.

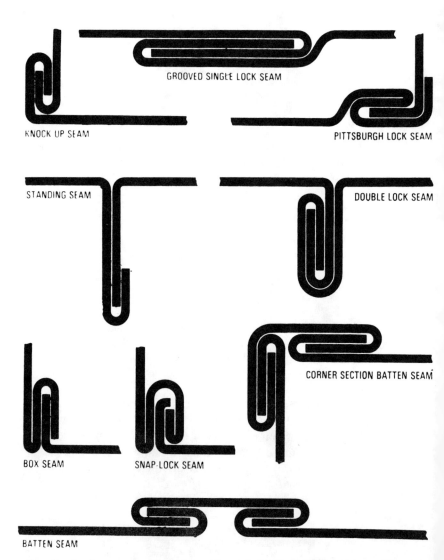

FIG. 8.   Lock seam configurations. (Reproduced from *Sheet Metal Industries*, **54**(2), 108–20, 1977; with permission, Fuel and Metallurgical Journals Ltd.)

3.6. Special Types

In Scheme 2 the special types mentioned are seaming and brazing, but, in practice many special types are available. A number of seaming types are shown in Fig. 8. Seaming is frequently applied in silos.

4. PRINCIPLES OF DESIGN

As mentioned in the introduction and in Scheme 1, there are two types of requirements. When on the basis of all relevant criteria the optimal type of fastener is selected, the number of fasteners is determined by the structural requirements.

For the design of a connection, an engineer has to compare two quantities:

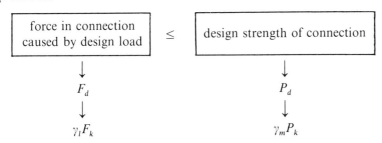

where F_k = force in connection caused by characteristic load,
   P_k = characteristic strength of connection,
   γ_l = appropriate load factor, and
   γ_m = appropriate material factor.

The forces in connections are dependent on:

—loads on the jointed elements;
—stiffness of the jointed elements;
—stiffness and deformation capacity (see Section 4.1) of the fastenings.

In Section 4.2 the item of forces in connections will be treated more in detail.

The strength of connections is dependent on:

—type of fastener;
—properties of jointed elements (thickness, yield stress).

In Section 5 the strength item will be treated in detail.

4.1. Mechanical Properties of Fastenings (Qualitative)

Thick-walled structural steel and also the light gauge steel sheet clearly possess such properties as strength, stiffness and deformation capacity (Fig. 9). Therefore, this material is suitable for use in structures. Ancillary parts of structures, e.g. fastenings, ought to have the same properties (Fig. 10). Evidently, this applies to their strength and stiffness. It is less well known, however, that their deformation capacity should also meet certain requirements.

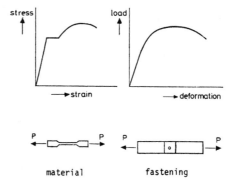

FIG. 9.   Structural behaviour. (Reproduced from *Sheet Metal Industries*, **54**(2), 108–20, 1977; with permission, Fuel and Metallurgical Journals Ltd.)

4.1.1. Strength
For each type of fastening, a characteristic strength can be determined by theoretical or experimental research. This strength can be influenced by the choice of the section, and by the type of fastener.

4.1.2. Stiffness
Stiffness of a connection is important because it determines the stiffness of the whole structure or of its components. Moreover stiffness can also influence the forces in a connection. This is, for instance, the case for connections of lateral bracings to purlins and bracings in nonsway frames. The stiffness of a connection also determines the distribution of the loads.

4.1.3. Deformation Capacity
The deformation capacity of a connection is important. A connection with no deformation capacity can cause a brittle fracture of a structure or element. This primarily applies to continuous construction, where such

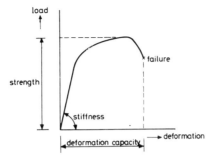

FIG. 10.   Definition of strength, stiffness and deformation capacity. (Reproduced from *Sheet Metal Industries*, **54**(2), 108–20, 1977; with permission, Fuel and Metallurgical Journals Ltd.)

influences as settlings and fluctuating temperatures are normally not included in a design calculation. Local overloading can be eliminated if the connection can deform sufficiently.

In the case of simply supported structures, the deformation capacity of the separate fasteners can be important if more fasteners are used. Figure 11 gives two examples which demonstrate the importance of this requirement:

1.   A failure mode with little strain capacity has disadvantages if many fasteners have been placed in a row in the direction of load. In a calculation, the force is divided into equal parts and applied to each fastener. Theoretically this is not correct. However, this assumption may be used if plastic redistribution can take place.

2.   The same requirement is necessary for connections in trusses which are calculated as trusses with pin-ended joints. It is well-known that secondary stresses are always introduced. A simplified design method is therefore permitted if the connections can deform plastically in order to limit the influence of the secondary stresses.

4.2. Forces in Connections

The forces in connections are dependent on the external loading on the structure and on the static properties of the structure. In general the external loads have to be taken in accordance with national standards. The loads will cause shear, tension or a combination of these in the fastenings. The philosophy in determining these forces in sections or profiled sheeting will be treated in Sections 4.2.1 and 4.2.2.

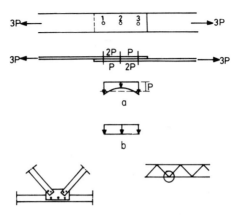

FIG. 11. Clarification of the necessity of deformation capacity. (a) Theoretical
distribution (without plastic deformations); (b) calculation.

4.2.1. Forces in Connections in Sections

Distinction has to be made between continuous construction and simply
supported structures. For simply supported structures the force on a
connection follows from equilibrium (Fig. 12). The distribution of the
forces over the fastenings depends in principal on the stiffness per fastening.
A simplified calculation leads to the following conservative maximum force
per fastening:

$$F = \sqrt{\left(\frac{Pa}{2h}\right)^2 + \left(\frac{P}{6}\right)^2}$$

where F is the force per fastening.

For continuous construction the design philosophy will be treated for the
example shown in Fig. 13. When designing the splice connection the
procedure now described has to be followed:

—Determine the rotation ϕ_c of the connection as a function of the
   stiffness per fastening and a moment $M \to \phi_c = f$ (M, fastening
   stiffness).
—Determine the inclinations ϕ_b of the beam over the support as a
   function of the support moment, beam stiffness and load (see
   Fig. 14) $\to \phi_b = f$ (M, l, EI, q).
—Compatibility requirement:

$$\phi_c = \phi_b \to M_{support} \to \text{force per fastening}$$

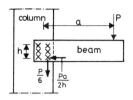

FIG. 12.   Force on connection in statical determined structure. Moment on connection = $P \times a$. Shear force on connection = P.

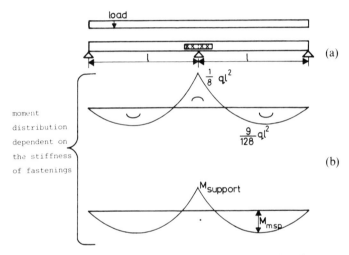

FIG. 13.   Beam on three supports with splice. (a) For a fastening infinitely stiff. (b) For a flexible fastening: $M_{support} < \frac{1}{8}ql^2$; $\frac{9}{128} < (M_{m.sp}/ql^2) < \frac{1}{8}$.

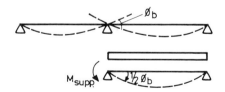

FIG. 14.   Inclination of beam over support.

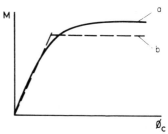

FIG. 15. Relation between moment and rotation of a connection. (a) Tested
relation; (b) relation for use in calculations.

Remarks:

(a) In the case of the bi-linear behaviour of the connection (Fig. 15), the horizontal part of the curve may be used; then the distribution of forces simply follows from equilibrium. The compatibility requirement need not be met; however, proof is required that the connection possesses sufficient deformation capacity.

(b) In the case of little stiffness of the connection, the connection shall be regarded as a hinge (two simply supported beams). The connection shall then possess sufficient deformation capacity to follow the inclination of both beams.

As an example of the determination of forces three types of structures in thin-walled sections will be examined:

(a) Composition of a bending member from single sections with connections loaded in shear (see Fig. 16): Shear force in fastenings in a cross-section A:

$$S_A = \frac{aVZ}{I}$$

where S_A = sum of shear forces in both fastenings in a cross-section A; a = distance between fasteners in span direction; V = vertical shear force in A equal to $\frac{1}{2}qL$; Z = area of the part which will shear, multiplied with the distance of the centroid of the area which will shear to the neutral axis of the composite beam; and I = moment of inertia of the composite beam. The calculation method shown gives an upper-limit of the shear force in the fastenings A. In reality some slip in the fastenings will occur. This causes a smaller section modulus and moment of inertia of the composite beam and reduces the forces on the fastenings.

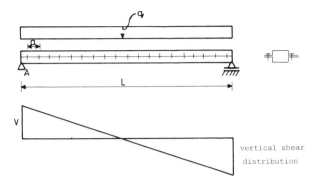

FIG. 16.   Shear forces in a composite beam.

(b) Composition of a bending member from single sections with connections loaded in tension: Figure 17 shows the cross-section of two C-sections connected to each other. The method of calculating the forces in the fastenings is described in the figure.

(c) Secondary forces in connections. Care should be taken, by suitable detailing, that second order effects caused by deformation of thin-walled sections will not generate impermissible extra forces in the fastenings. This is illustrated by the example given in Fig. 18.

4.2.2. Forces in Connections in Profiled Sheeting

For connections in profiled sheeting distinction will be made between primary and secondary forces:

(a) Primary forces—forces which are directly caused by the load;

(b) Secondary forces—forces which are indirectly caused by the load and which may be neglected in the presence of sufficient deformation capacity (see Section 4.3).

Scheme 3 gives a survey of primary and secondary forces in profiled sheeting.

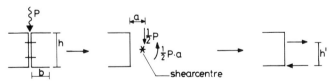

FIG. 17.   Cross-section of two C-sections connected to each other. Tension force in fastener $= \frac{1}{2}Pa/h$. For a C-section: $a = 3b^2/(6b + h)$.

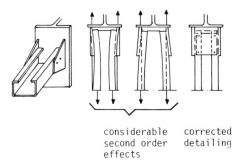

considerable   corrected
second order   detailing
effects

FIG. 18.   Illustration of the influence of detailing of a connection on the
deformations.

SHEAR

   Primary
                   the dead weight of steel sheets
                   diaphragm action when the diaphragm is used
                     deliberately
                   variation of the temperature of steel sheets

   Secondary

- the dead weight of steel sheets
- *rotation of the eccentric (with regard to the neutral axis of the steel sheet) fastened sheet ends and the membrane-action of the sheet*
- *diaphragm action which is not used structurally*

TENSION   Primary   *'uplift' loads*
         Secondary   *prying forces*

SCHEME 3.   Survey of forces in connections in profiled sheeting.

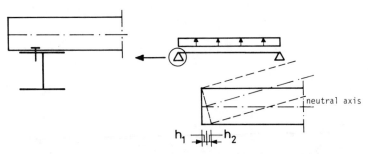

FIG. 19.   Support of sheet in detail; h_1 = deflection caused by difference in length between straight and curved neutral axis; h_2 = deflection caused by distance between neutral axis and flange through which it is fastened.

4.2.2.1. Shear forces in connections in profiled sheeting. These are as follows:

(a) The dead weight of steel sheets; for instance, the weight of wall or façade elements;

(b) The diaphragm action, when the diaphragm is used deliberately; for instance, in the absence of a wind bracing, stability support for beams or columns; references 11 and 12 give calculation methods for these aspects;

(c) Variation of the temperature of the steel sheets; with sufficient deformation capacity the shear forces will be small and may be neglected;

(d) Rotation of the eccentric fastened sheet ends and the membrane-action of the sheet (see Fig. 19); in the presence of sufficient deformation capacity the fastening will not fail;

(e) Diaphragm action which is not used structurally; this is the case when a sheeting or cladding is only used as an outer skin; it is then necessary for the skin to follow the deformations of the sub-structure; this is possible when the diaphragm (especially the fastenings) possesses sufficient deformation capacity.

4.2.2.2. Tension forces in connections in profiled sheeting. Tension forces in fastenings will be caused mainly by loads perpendicular to the plane of the steel sheets. For the determination of the required strength and stiffness of the sheets a simply supported statical system is assumed (Fig. 20). In reality

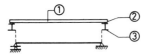

FIG. 20.   Statical system to determine the required strength and stiffness of a steel sheet. 1. Steel sheet. 2. Fastener. 3. Support.

the sheets are to some extent restrained at the supports; but for the design of the sheets it is safe to neglect the restraining effect. For the design of the fastenings simplification to a simply supported statical system gives forces which are too small. In the absence of sufficient deformation capacity the fastening can fail at a premature stage.

The action of the internal forces in a fastening is as follows (Fig. 21): Due to bending of the steel sheet a compression force will act on the supports at

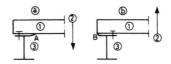

FIG. 21.   Detail of the support of a steel sheet. (a) Downward load, (b) uplift load.
1. Steel sheet. 2. Load (see text for key to A and B). 3. Support.

point A or B. This causes an accidental fixing moment for the steel sheets, which generates an extra tension force in the fastener; the prying-force. The value of the prying-force depends on:

1. the stiffness of the sheets in relation to the span;
2. the flexibility of the sheets near the fastener;
3. the diameter of the head of the fastener or the diameter and stiffness of the washer;
4. the distance between the fastener and the contact points A or B;
5. the torsional rigidity of the support.

When sufficient deformation capacity is available the required rotation can take place and the first order reaction can proceed.

4.3. Deformation Capacity
Principally, the deformation capacity of a fastening has to be determined by tests. Reference 12 gives criteria for the required deformation capacity. In Section 5 formulae are given for the design strength of fastenings. Also limits are presented that guarantee sufficient deformation capacity.

5.   DESIGN STRENGTH OF FASTENINGS

5.1. General
In reference 13 standard testing procedures are presented to determine characteristic values for fastenings. For design purposes various formulae have been derived based on experiments (see references 7, 12 and Swedish Codes); such formulae are presented in Section 5.4. It should be borne in mind that the design formulae for fastenings are as a consequence of their nature in general conservative. Design values determined by tests will give more realistic values.

The formulae given are valid only for fastenings in which the fastened parts are in direct contact with each other. That means that only small bending moments are acting on the fasteners and the formulae are not valid for the crest fastening of profiled sheeting or the fastening of sandwich elements for example.

In the fastening of profiled sheeting to the substructure different fastening types will appear because of overlap. The four most important are shown in Fig. 22.

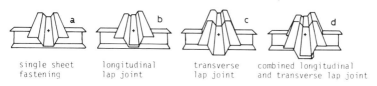

| single sheet | longitudinal | transverse | combined longitudinal |
| fastening | lap joint | lap joint | and transverse lap joint |

FIG. 22.   Fastening types in profiled sheeting.

To simplify the design of the fastenings between profiled sheeting and substructures no distinction is made between the different fastening types. For this reason the most unfavourable fastening type for the different failure modes has to be taken for the calculation. The lowest strength for sheet failure will be reached for fastening type 'a' in Fig. 22. Fastening type 'd' gives the smallest deformations and also, for powder actuated fasteners and screws with short penetration depth, the lowest pull-out strength. Tests to determine the pull-out strength and the deformation capacity therefore have to be carried out using test specimens as in Fig. 23.

5.2. Failure Modes of Fastenings

The strength and flexibility of fastenings is dependent on the failure mode of the fastening. Sections 5.2.1 and 5.2.2 survey the possible failure modes for shear and tension, respectively.

5.2.1. Failure Modes of Fastenings Loaded in Shear
(a) Shear of fastener:

This may occur when the sheet is thick with reference to fastener diameter, or when an unsuitable fastener is used.

(b) Crushing of fastener:

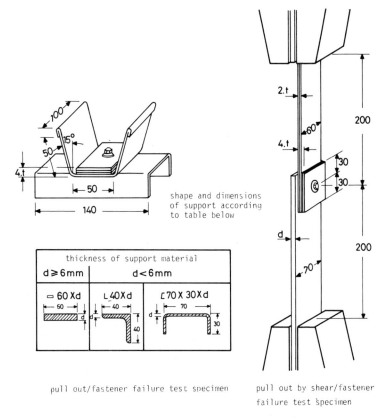

pull out/fastener failure test specimen   pull out by shear/fastener
                                                  failure test specimen

FIG. 23.   Test set up to determine pull out strength and deformation capacity.

This may occur with hollow fasteners, and in combination with tilting an
yield in bearing.

(c) *Tilting and pull-out of fastener* (*inclination failure*):

This is the normal mode of failure in thin sheet to thin sheet fastening wit
either the threads or the site formed rivet head pulling out of the lowe
sheet. It may occur in combination with yield of both sheets in bearing, an
in conjunction with considerable sheet distortion.

(*d*) *Yield in tearing* (*tearing of sheet*):
(i) Yield of thinner sheet only.

(ii) Yield of both sheets.

(*e*) *End failure:*

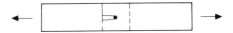

This will not occur if recommended end distances are adhered to.

(*f*) *Failure of the net cross-section.*

5.2.2. *Failure Modes of Fastenings Loaded in Tension*
(*a*) *Tension failure of fastener:*

This may occur when the sheet is thick with reference to the fastener, or when an unsuitable fastener is used.

(*b*) *Pull out:*

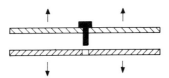

This may occur when the support member is insufficiently thick, or when there is insufficient anchorage of fastener.

(*c*) *Pull over:*

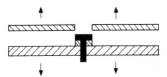

(*d*) *Pull through:*

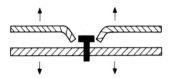

This may be accompanied by washer distortion.

(*e*) *Gross distortion of sheeting:*

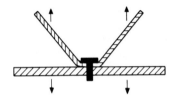

Permanent gross profile distortion may be considered to constitute unserviceability of the structure.

5.3. Some Test Results
It is of course impossible to show all the test results. Many thousands of tests have been carried out in various research programmes. Some results of the test programme on bolted connections will be discussed. In addition some results of the tests on screws will be considered.

5.3.1. Tests on Bolted Fastenings
This research programme has been carried out by the University of Strathclyde, Glasgow, UK (see reference 15). Figure 24 shows the connections investigated. Three thicknesses of sheet steel were used, namely 1, 3 and 5 mm. Three sizes of high-strength bolts were used with diameters 5, 10 and 16 mm, respectively. The influence of the bolt tightness was investigated. It has been found that the connection strength is independent of the bolt torque. In the tests on flat sheet connections as well as the connections of channel sections the same four modes of failure were found. These modes of failure are summarised in Fig. 25 which also gives the design rules derived from the tests.

5.3.1.1. Failure 1: Edge failure (Fig. 26). The bolts remained perpendicular to the applied load direction. The shear planes were parallel and a distance apart equal to the bolt diameter. A slight amount of oblique

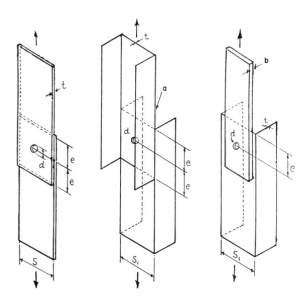

FIG. 24. Test specimens for bolted connections. (a) One and two bolt connections tested for this case. (b) Width = 16 mm. (Reproduced from *Sheet Metal Industries*, **54**(2), 108–20, 1977; with permission, Fuel and Metallurgical Journals Ltd.)

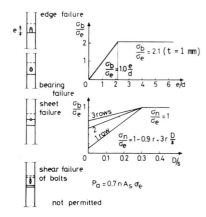

FIG. 25. Bolted fastenings. (Reproduced from *Sheet Metal Industries*, **54**(2), 108–20, 1977; with permission, Fuel and Metallurgical Journals Ltd.)

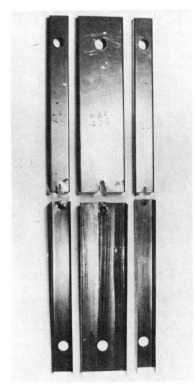

FIG. 26. The results of edge failure.   FIG. 27.   The results of bearing failure.

tearing occurred at a point near complete failure of the connections. It seems obvious that for this failure mode the ultimate force is dependent of the edge distance c_1.

Design rule:

$$\frac{\text{design bearing stress}}{\text{yield stress}} = 1 \cdot 0 \frac{\text{edge distance}}{\text{diameter of bolt}}$$

5.3.1.2. Failure 2: Bearing failure (Fig. 27). A bearing failure of the sheet material with a shearing–tearing action along two distinctly inclined planes, caused the sheet material to pile up in front of this bolt. In some of the tests it was found that the bolt turned parallel to the load direction and eventually pulled through the material. For this failure mode the ultimate load is not dependent of the edge distance.

Design rule:

$$\frac{\text{design bearing stress}}{\text{yield stress}} = \text{function of } t \qquad (\text{see Section } 5.4)$$

5.3.1.3. Failure 3: Sheet failure (see Fig. 28). Transverse tearing takes place at the net section and is at right angles to the load direction. In the case of C-sections it changes to a more oblique angle as it proceeds across the flanges to complete failure. If the spacing of the bolts is close the ultimate stress over the net section is equal to the yield stress. If the strap is wide or the spacing in a transverse direction is large, because of high concentrated forces, failure at the hole occurs before the whole net section yields.

This phenomenon is dependent on the ratio between the force in the bolt

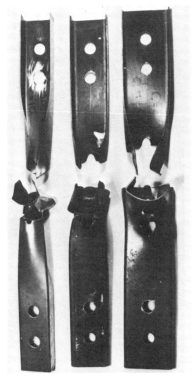

FIG. 28.   The results of sheet failure.   FIG. 29.   The results of shear failure of the bolt.

and the force in the sheet. The more bolts in a longitudinal row the higher the ultimate net stress in this case.

5.3.1.4. Failure 4: Shear failure of the bolt (Fig. 29). Failure of the bolt occurs at a mean shear stress of about 0·7 times the ultimate stress of the bolt material. In terms of design strength the following formula is found:

$$P_a = 0·7 n A_s \sigma_e$$

where n = number of shear planes. This failure type occurs with relatively small deformations and is therefore not permitted. It is required that P_a is at least 1·3 times the smallest value calculated for the other failure modes.

5.3.2. Tests on Screwed Fastenings

This part of the programme has been carried out at IBBC-TNO in The Netherlands (reference 16). As shown in Fig. 30, almost the same failure modes occur as with bolted connections. However, there is one additional failure mode, namely the inclination failure. What happens in this case will be clear from the photographs shown in Fig. 31.

Figure 32 is a characteristic load–deformation diagram for inclination failure. The ultimate load is strongly dependent on the thickness of the sheet in which it is screwed, because this sheet forms the nut. For $t_1/t = 1$, Fig. 30 gives the design formulae. For $t_1/t > 1$ the design strength is related

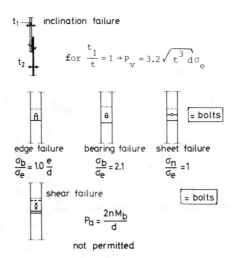

FIG. 30.   Fastenings with screws. (Reproduced from *Sheet Metal Industries*, **54**(2), 108–20, 1977; with permission, Fuel and Metallurgical Journals Ltd.)

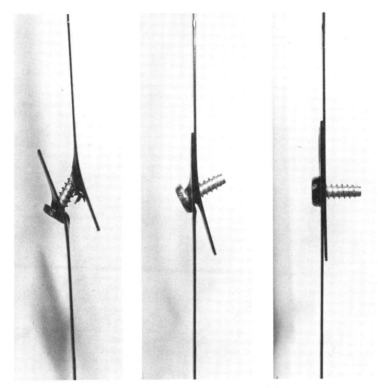

FIG. 31. The stages of inclination failure.

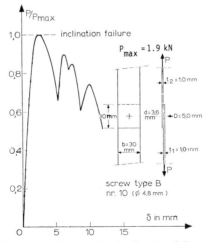

FIG. 32. Load–deformation diagram for inclination failure. (Reproduced from *Sheet Metal Industries*, **54**(2), 108–20, 1977; with permission, Fuel and Metallurgical Journals Ltd.)

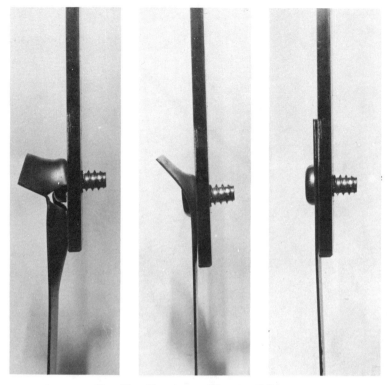

FIG. 33.   The stages of bearing failure.

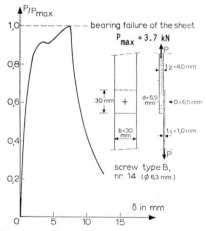

FIG. 34.   Load–deformation diagram for bearing failure. (Reproduced from *Sheet Metal Industries*, **54**(2), 108–20, 1977; with permission, Fuel and Metallurgical Journals Ltd.)

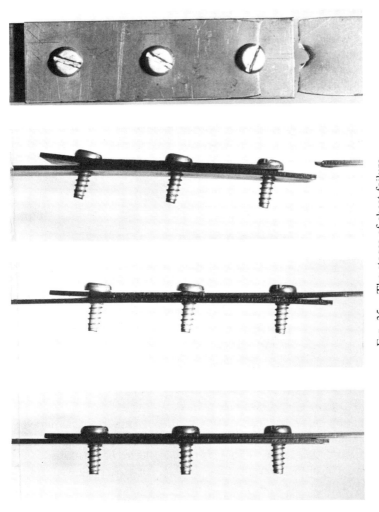

FIG. 35.   The stages of sheet failure.

to the bearing strength. As a simplification linear interpolation is recommended between the inclination value at $t_1/t = 1$ and the bearing strength at $t_1/t = 2·5$.
The photographs in Fig. 33 provide an impression of the bearing failure mode. The head of the screw is almost drawn through the sheet material. A measured load–deformation diagram is shown in Fig. 34.
Sheet failure is shown in Fig. 35 with the load–deformation diagram in Fig. 36. The mode of shear failure of the screw is shown in Fig. 37.

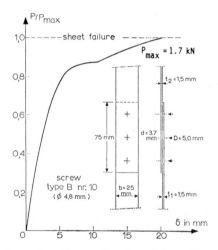

FIG. 36.   Load–deformation diagram for sheet failure. (Reproduced from *Sheet Metal Industries*, **54**(2), 108–20, 1977; with permission, Fuel and Metallurgical Journals Ltd.)

The load–deformation diagram shown in Fig. 38 illustrates the small deformation capacity. Therefore, this failure mode is not permitted (see Fig. 30). Because the ultimate shear stress of the screw material is not known, the design values of P_a is expressed in terms of the ultimate torque moment, M_b, of the screw. The value of M_b for a particular screw type is given in national standards.

5.3.3. Tests on Complete Connections
It is appropriate to consider how the test results of standard test specimens can be used to calculate the properties of complete connections. From the actual measured load–deformation diagram a bi-linear design diagram is

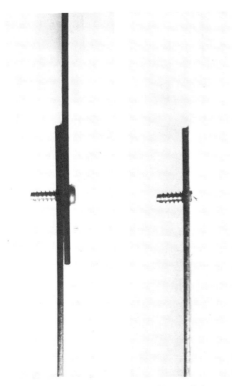

FIG. 37.   The stages of shear failure of the screw.

derived as shown in Fig. 39. The values of stiffness and strength are now known. For the T-connection loaded in bending and shear as shown in Fig. 40 a moment–curvature relation can be calculated (Fig. 39). This type of calculation was effected for a T-connection between cold-formed top-hat sections and channel sections. Figure 40 gives an impression of the connection. In Fig. 41, the results for one case are shown. It can be seen that under working load the agreement between calculation and the test result is good. Furthermore, there is a relatively high reserve of strength. This is caused by redistribution of forces between the fasteners after reaching the ultimate load in the most heavily loaded fastener. In this case the failure mode is inclination failure (Fig. 42).

Figures 43 and 44 show other modes of failure, namely bearing failure and shear failure, respectively.

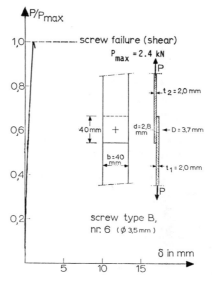

FIG. 38.   Load–deformation diagram for shear failure of the screw. (Reproduced from *Sheet Metal Industries*, **54**(2), 108–20, 1977; with permission, Fuel and Metallurgical Journals Ltd.)

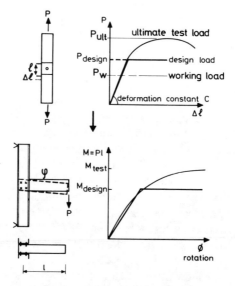

FIG. 39.   Bi-linear design diagram.

FIG. 40.   T-connection loaded in bending and shear.

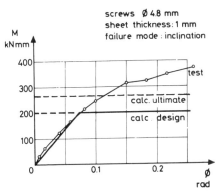

FIG. 41.   Impression of a T-connection between cold-formed, top-hat sections and channel sections. (Reproduced from *Sheet Metal Industries*, **54**(2), 108–20, 1977; with permission, Fuel and Metallurgical Journals Ltd.)

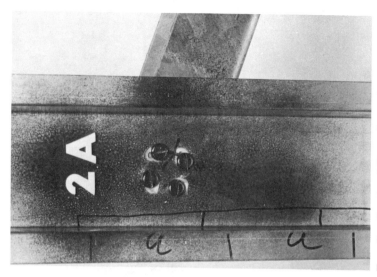

Fig. 43.   Bearing failure of complete connection.

Fig. 42.   Inclination failure of complete connection.

FIG. 44.   Shear failure of complete connection.

5.4. Formulae for the Design Strength of Fastenings

5.4.1. Design Strength for Mechanical Fastenings Loaded in Shear
Table 1 gives a survey of design formulae for the various failure modes of
fastenings. The design strength of a fastening is the smallest value resulting
from the formulae for a given type of fastener. In Table 1 the following
symbols are used:

d = diameter of fastener
P = design strength (subscript depends on failure mode)
σ_e = design value for the yield stress of the sheet material
A_n = net cross-sectional area of the plate material
r = force transmitted by the bolt or bolts at the section considered,
    divided by the tension force in the member at that section
D = diameter of the hole

TABLE 1
DESIGN SHEAR STRENGTH FOR FASTENINGS

Fastener type	Failure mode with design shear strength (in N)					Application limits for formulas*
	Tilting + hole bearing, $t < t_1$	Hole bearing	Shear of section	Yield of net section	Shear of fastener	
Blind rivets	$t = t_1$: $P_t = 3.6\sqrt{t^3}\,d\sigma_e$ $P_t = 2.1\,t\,d\sigma_e$ $t_1 \geq 2.5t$: $1 \leq \dfrac{t_1}{t} \leq 2.5$: linear interpolation	$P_b = 2.1\,t\,d\sigma_e$	$\left\{\begin{array}{l}\text{Not relevant}\\ \text{when } c_1 \geq 3d\end{array}\right.$	$P_n = A_n\sigma_e$	$\left\{\begin{array}{l}\text{Characteristic}\\ \text{shear strength}\\ \text{of fastener } 1.5\\ \text{times larger}\\ \text{than other}\\ \text{failure modes}\end{array}\right.$	$\left\{\begin{array}{l} c_1 \geq 3d \\ c_2 \geq 3d \\ e_2 \geq 3d \\ e_1 \geq 1.5d \\ 2.6\text{ mm} \leq d \leq 6.4\text{ mm} \end{array}\right.$
Bolts with nuts	Not relevant	$t \leq 1$ mm: $P_b = 2.1\,t\,d\sigma_e$ $P_b = \begin{cases} 1\text{ mm} < t < 3\text{ mm}\ \text{and}\ \dfrac{c_1}{d} \leq 6 \\[4pt] \left\{2.6 - 0.5t + 0.9(t-1)\ln\dfrac{c_1}{d}\right\} t\,d\sigma_e \\[6pt] 1\text{ mm} < t < 3\text{ mm}\ \text{and}\ \dfrac{c_1}{d} > 6 \\[4pt] P_b = (1.0 + 1.1t)\,d\sigma_e \\[6pt] t \geq 3\text{ mm}\ \text{and}\ \dfrac{c_1}{d} \leq 6 \\[4pt] P_b = \left(1.1 + 1.8\ln\dfrac{c_1}{d}\right) d\,t\sigma_e \\[6pt] t \geq 3\text{ mm}\ \text{and}\ \dfrac{c_1}{d} > 6 \\[4pt] P_b = 4.3\,t\,d\sigma_e \end{cases}$	$P_s = t\,c_1\,\sigma_e$	$\left\{\begin{array}{l} P_n = A_n\sigma_m < A_n\sigma_e \\[4pt] \sigma_n = \left(1 - 0.9r + 3r\,\dfrac{D}{s}\right)\sigma_e \end{array}\right.$	$\left\{\begin{array}{l} P_f = 0.7\,n\,A_s\sigma_b, \\ P_f \text{ has to be} \\ 1.3 \text{ times} \\ \text{larger than} \\ \text{other failure} \\ \text{modes} \end{array}\right.$	$\left\{\begin{array}{l} c_1 \geq 1.5d \\ c_2 \geq 4d \\ e_2 \geq 4d \\ e_1 \geq 1.5d \\ \text{M6-M16} \\ 8.8 \text{ or } 10.9 \end{array}\right.$

Screws	$t = t_1$: $\quad P_t = 3 \cdot 2\sqrt{t^3 d\sigma_e}$; $\quad P_t = 2 \cdot 1 t d\sigma_e$ $t_1 \geq 2 \cdot 5t$: $1 \leq \dfrac{t_1}{t} \leq 2 \cdot 5$: linear interpolation	$P_b = 2 \cdot 1 t d\sigma_e$	$\begin{cases}\text{Not relevant}\\\text{when } c_1 \geq 3d\end{cases}$	$P_n = A_n \sigma_e$	$\left\{\begin{array}{l}\text{Characteristic}\\\text{shear strength}\\\text{of fastener } 1 \cdot 5\\\text{times larger}\\\text{than other}\\\text{failure modes}\end{array}\right.$ $\quad\left\{\begin{array}{l}c_1 \geq 3d\\c_2 \geq 3d\\e_2 \geq 3d\\e_1 \geq 1 \cdot 5d\\3 \cdot 0\,\text{mm} \leq d \leq 8 \cdot 0\,\text{mm}\end{array}\right.$
Powder actuated fasteners	Not relevant	$P_b = 3 \cdot 2 t d\sigma_e$	$\begin{cases}\text{Not relevant}\\\text{when } c_1 \geq 4 \cdot 5d\end{cases}$	$\begin{cases}\text{Not relevant in}\\\text{sheeting}\end{cases}$	$\left\{\begin{array}{l}\text{The pull-out}\\\text{load due to}\\\text{shear must}\\\text{exceed } 1 \cdot 5\\\text{times } P_b\end{array}\right.$ $\quad\left\{\begin{array}{l}c_1 \geq 4 \cdot 5d\\c_2 \geq 4 \cdot 5d\\e_2 \geq 4 \cdot 5d\\e_1 \geq 4 \cdot 5d\\3 \cdot 7\,\text{mm} \leq d \leq 6\,\text{mm}\\\text{base material}\\t_1 \geq 6\,\text{mm}\end{array}\right.$

* These limits are only limits for the application of the formulas and not limits for the use of the fasteners. The limits are imposed by the extent of the relevant research.

s = spacing of bolts perpendicular to the line of stress; in the case of a single bolt, s = width of plate

n = number of shear planes in a bolt

A_s = stress cross-sectional area of a bolt

σ_b = design value of the yield stress of the bolt material

5.4.2. Design Strength for Mechanical Fastenings Loaded in Tension

Table 2 gives a survey of design formulae for the various failure modes of fastenings. These formulae are specially derived for sheeting; therefore only screws and powder actuated fasteners are treated. These are the fasteners normally used for this application. In Table 2 the following symbols are used:

d = diameter of fastener

P = design strength (subscript depends on failure mode)

σ_e = design value for the yield stress of the sheet material

σ_{ec} = design value for the yield stress of the base material

5.4.3. Design Strength for Welded Connections

The following types of connections are discussed:

—spot welded sheet steel, resistance and fusion welded;
—lap fillet welds, fusion welded.

The formulae and data apply for welds in parent materials with thickness equal to or smaller than 4 mm. For welds in parent materials over 4 mm thickness the normal welding specifications apply.

5.4.3.1. Design requirements. These are as follows:

(a) Spot welded sheet steel connections:

The shear force per spot weld, used in a lap joint, due to the design load should not exceed the design strength indicated in Table 3. In Table 3 the following symbols are used:

P = design strength of a single spot weld (subscript depends on failure mode)

d_s = diameter of spot weld; the value of d_s may be taken as:

for fusion welding: $d_s = 5 + 0.5t$
for resistance welding: $d_s = 5\sqrt{t}$

(this value of d_s shall be controlled by standard shear tests as discussed later in this section)

t = thickness of thinnest connected member

σ_e = design values for the yield stress of the material

e = edge distance of spot weld

A_n = net cross-sectional area of plate material

TABLE 2
DESIGN TENSION STRENGTH FOR FASTENINGS

Fastener type	Failure mode			Application limit for formulae
	Pull through pull over	Pull out	Tensile failure	
Screws	$P_p = 15t\sigma_e$, or $P_p = 7.5t\sigma_e$. (See footnotes a, b, c)	$P_0 = 0.65t_c\sigma_{ec}$. (See footnote d)	Characteristic tensile strength of fastener larger than P_p or P_0. (See footnote e)	0.5 mm < t < 1.5 mm, $t_c > 0.9$ mm
Powder actuated fastener	$P_p = 15t\sigma_e$, or $P_p = 7.5t\sigma_e$. (See footnotes a, b, c)	Characteristic pull out strength larger than P_p. (See footnote e)	Not relevant in sheeting	0.5 mm < t < 1.5 mm, $t_c > 6$ mm

a The formula $P_p = 15t\sigma_e$ is valid for static loads. The formula $P_p = 7.5t\sigma_e$ is valid for repeated loads with a spectrum similar to wind.

b The failure mode 'gross distortion' is not covered by the given formula. For fastenings through flanges with a width smaller than 150 mm, the local deformation of the flange under working load is in most cases in the elastic range.

c It is assumed that load is applied centrally and that the head of the fastener or washer, if any, has a diameter of at least 14 mm and has sufficient rigidity to prevent it being deformed appreciably. When attachment is at a quarter point, the design value is $0.9P_p$, and when it is at both quarter points, it is $0.7P_p$.

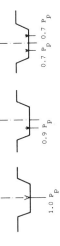

1.0 P_p   0.9 P_p   0.7 P_p 0.7 P_p

d When $P_0 \ll P_p$, it has to be proved that the deformation capacity is sufficient.

e For more than one sheet layer the tensile strength of the fastener should be more than one times P_p or P_0. The factor depends on the number of sheet layers.

TABLE 3
DESIGN STRENGTH PER SPOT WELD

Failure modes				Remarks
Shear of spot weld	*Tearing + bearing at contour weld*	*Edge failure (shear section)*	*Net section*	

$P_s = \dfrac{\pi}{4} d_s^2 \sigma_e$ $P_t = 3 \cdot 5 t d_s \sigma_e$ $P_e = 1 \cdot 4 t e \sigma_e$ $P_n = A_n \sigma_e$ Constructional details according to Section 5.4.3.2

Necessary condition to be fulfilled: $P_s \geq 1 \cdot 25 \times \begin{bmatrix} P_t \\ P_e \\ P_n \end{bmatrix}$

The value of d_s, depending on the weld procedure, will be controlled by shear tests of single lap connections according to Fig. 45. The characteristic shear strength, R, will be determined by statistical evaluation of the test results.

The requirement for a sound weld is that $R \geq P_s$.

(b) *Lap fillet welds:*
The shear force of lap fillet welds, due to the design load, should not exceed the design strength given by the following formulas:

—End fillet welds: $P = t l_w \sigma_e (1 - 0 \cdot 3(l_w/b))$ per weld
—Side fillet welds: $P = t 2 l_w \sigma_e (0 \cdot 9 - 0 \cdot 45(l_w/b))$ per two welds, provided $l_w \leq b$

l_w = effective length of weld; for the purpose of strength calculation the effective length of a fillet weld will be the theoretical length without a deduction for head and/or end craters; weld lengths shorter than eight times the thickness of the thinnest connected sheet will be ignored for transmission of forces;
b = width of specimen;

The remaining symbols are similar to those in Section 5.4.3.1(a).

In a joint where a combination of end fillet welds and side fillet welds is applied, the total design strength will be the sum of the strengths of the individual welds.

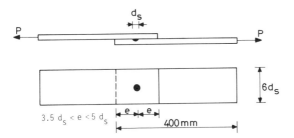

FIG. 45.   Test specimen shear failure tests.

Designing lap filled welded connections loaded in shear (Fig. 46), it should be noted that the deformation capacity of both end and side fillet welds is relatively small. Nevertheless the deformation capacity is adequate to allow for addition of the strength of individual end filled welds and side fillet welds in a joint. However sufficient deformation capacity of the whole joint will only be attained when yielding can occur in the material outside the joint (a full strength connection).

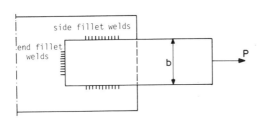

FIG. 46.   Lap fillet welds of specimen loaded in shear.

(c) *Fillet welds in T-joints:*
The plate elements of cold formed thin-walled sections have little rigidity when loaded transverse to the plate surface. (This loading situation occurs in T-joints.) As a consequence the stress distribution is highly unequal. This is illustrated in Fig. 47 for a T-joint of a strip and a cold-formed rectangular hollow section. When the weld is designed for full strength, yielding of the strip will cause redistribution of stresses. However, it is quite possible that tearing starts at the edges before the full yield strength of the strip is attained. In design this effect may be included by introducing an effective

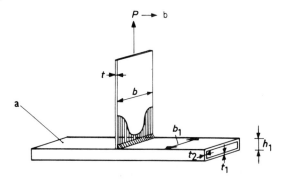

FIG. 47.   T-joint, section and strip. (a) Cold-formed rectangular hollow section:
$100 \times 40 \times 3$ mm; $t = 1\cdot5$ mm; $t_1 = t_2 = 2\cdot8$ mm; $b_1 = 100$ mm; $h_1 = 40$ mm;
$\sigma_u = 320$ N mm^{-2}. (b) Failure load, P_u.

length b_t for the weld. As an example, some test results for the connection of
Fig. 47 are presented:

$$b = 30 \text{ mm} \qquad P_u = 14\cdot7 \text{ kN} \qquad \frac{b_t}{b} = 1\cdot0$$

$$b = 60 \text{ mm} \qquad P_u = 25\cdot7 \text{ kN} \qquad \frac{b_t}{b} = 0\cdot89$$

$$b = 100 \text{ mm} \qquad P_u = 30\cdot0 \text{ kN} \qquad \frac{b_t}{b} = 0\cdot65$$

At this moment no suitable design formulae for the determination of b_t are
available. Effective length formulae for thick specimens as included in some
welding specifications have proved to be very conservative for thin-walled
sections.

In the Dutch welding specification NEN 2062 the following formula is
given for the effective weld length b_t in a connection as in Fig. 47:

$$b_t = 5t_1 + 2t_2$$

According to this formula the effective length would be $b_t = (5 \times 2\cdot8) +$
$(2 \times 2\cdot8) = 19\cdot6$ mm. This is much less than the test result which gives
$b_t = 65$ mm.

Thus, at this moment the only reasonable way to determine b_t is by
means of testing. Research is under way to develop design formulae.

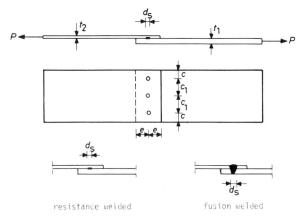

resistance welded   fusion welded

FIG. 48.   Spacing of spot welds.

5.4.3.2. Constructional details. The design formulae presented in Section 5.4.3.1 apply provided the following dimensional requirements are met:

(*a*) *Spot welding (Fig. 48):*
(i)   Spacing of spot welds. The edge distance from a spot weld to the edge of the connected part should not be less than $2d_s$ and not more than $6d_s$ in the direction of force and not more than $4d_s$ transverse to the direction of force. The centre to centre distance of two spot welds should not be less than $3d_s$ and not more than $8d_s$ in the direction of force and not more than $6d_s$ transverse to the direction of force.
(ii) Maximum thickness of sheet. The thickness of the thinnest sheet to be spot welded shall be not more than 3·0 mm.

(*b*) *Fillet welding (Fig. 49):*
The welding parameters shall be chosen so that the strength is governed by the thickness of the sheet. This will be achieved when the smallest cross-section of the weld is at least equal to the cross-section of the connected sheet.

FIG. 49.   Real weld thickness *s*.

The design formulae for fillet welded connections are based on failure of the plate material. For very thin material ($t < 2$ mm) this condition is always fulfilled; for thicker material ($2 < t < 4$ mm) care has to be taken to achieve at least a real weld thickness equal to the thickness of the thinnest connected member.

6. CONCLUDING REMARKS

This chapter is not intended to give solutions for all the structures which can be encountered in practice. Instead, it provides a design philosophy which is applied to a number of examples. The formulae given in Section 5.4 lead to conservative results because of the different makes of fasteners which are all included within the design formulae for each group. In certain cases, particularly for manufacturers of fasteners, it would be advantageous to determine more realistic values by carrying out tests (see reference 12) on the particular fastener in question.

REFERENCES

1. BAEHRE, R. and BERGGREN, L., Byggforskningen Hopfogning av Tunnväggiga stal-och aluminiumkonstruktioner 2. *Rapport R 30, Svensk Byggtjänst,* Stockholm, Sweden, 1971.
2. GROSSBERNDT, H. and KNIESE, A., Untersuchung über Querkraft- und Zugkraftbeanspruchungen sowie Folgerungen über kombinierte Beanspruchungen von Schraubenverbindungen bei Stahlprofilblech-Konstruktionen. *Der Stahlbau,* Heft 10 + 11, 1975.
3. KLEE, S. and SEEGER, T., Vorschlag zur vereinfachten Ermittlung von zulässigen Kräften für Befestigungen von Stahltrapezblechen. *Veröffentlichung des Instituts für Statik und Stahlbau der Technischen Hochschule,* Darmstadt, 1979.
4. STRNAD, M., Screwed connections in profiled sheeting. Paper of conference in Katowice, Poland, 30 May–2 June 1979.
5. NISSFOLK, B., Fatigue strength of joints in sheet metal panels. 2. Screwed and riveted connections, *Report D15: 1979, Swedish Council for Building Research,* Stockholm, 1979.
6. STARK, J. W. B., *Sheet Metal Industries,* **54**(2), February 1977.
7. STARK, J. W. B. and SOETENS, F., *Fifth International Speciality Conference on Cold-Formed Steel Structures,* Missouri-Rolla, November 1980.
8. STARK, J. W. B. and TOMÀ, A. W., *Fourth International Speciality Conference on Cold-Formed Steel Structures,* Missouri-Rolla, June 1978.
9. STARK, J. W. B. and TOMÀ, A. W., *International Conference at the University of Strathclyde,* Glasgow, April 1979.

10. TOMÀ, A. W., Fastening steel sheets for walls and roofs to steel structures, *Acier-Stahl-Steel*, March 1979.
11. ECCS, *European Recommendations for the Stressed Skin Design of Steel Structures*, ECCS Committee T7, March 1977.
12. ECCS, *European Recommendations for Connections in Thin-Walled Structural Steel Elements*, ECCS Committee T7, February 1981.
13. ECCS, *European Recommendations for the Testing of Connections in Profiled Sheeting and Other Light Gauge Steel Components*, ECCS Committee T7, May 1978.
14. PONSCHAB, H. F. and KIEHAS, D., Connections in cold-formed sections by high-strength bolts in galvanized condition, *Bericht Nr EAF-43/75 of VOEST-Alpine*, Austria.
15. LOUGHLAN, J., RHODES, J. and HARVEY, J. M., *Bolted Connections in Thin-Walled Steel Sections Using High Tensile Bolts*. University of Strathclyde, Glasgow.
16. TOMÀ, A. W., Connections in cold-rolled sections. Part 2: Experimental Research on screwed connections, *Report IBBC-TNO, order nr 05.3.11.250*, March 1975.

Chapter 6

ON THE FOUNDATIONS OF PLASTIC BUCKLING

Viggo Tvergaard

Department of Solid Mechanics,
Technical University of Denmark, Lyngby, Denmark

and

Alan Needleman

Division of Engineering, Brown University,
Rhode Island, USA

SUMMARY

Present theoretical knowledge on the influence of plasticity on structural buckling relies partly on the general bifurcation theory for elastic–plastic solids together with related asymptotic analyses, and partly on a number of numerical investigations of plastic post-buckling behaviour. The bifurcation and initial post-bifurcation analyses are discussed briefly and related to elastic post-buckling theory. An asymptotic analysis of imperfection sensitivity, based on neglecting elastic unloading, is illustrated by a detailed investigation of plastic columns. Finally, several results of numerical post-buckling analyses for plate and shell structures are discussed.

1. INTRODUCTION

The buckling behaviour of elastic–plastic structures under compressive loading is the result of a complex interaction between geometric and material non-linearities. The plastic material behaviour provides the material non-linearities, but the phenomenon of buckling itself arises from the geometric non-linearities. Therefore, a good understanding of linear elastic buckling behaviour provides an important basis for studies of structural buckling in the plastic range.

The present detailed understanding of elastic post-buckling behaviour relies on the general theory of elastic stability developed by Koiter[1] in 1945. This theory has explained imperfection-sensitivity to be the cause of the often large discrepancies between the classical bifurcation load and experimentally observed buckling loads. Many results concerning the elastic post-buckling behaviour and imperfection-sensitivity of columns, frame structures, plate structures and shell structures under various loading conditions have been reviewed by Hutchinson and Koiter,[2] and more recent results are reviewed by Tvergaard.[3]

The analysis of plastic column buckling dates back to the last century, but it was not until 1947 that the significance of column bifurcation at the tangent modulus load was explained by Shanley.[4] About ten years later Hill[5,6] developed a general theory of bifurcation and uniqueness in elastic–plastic solids, which now forms the basis of practically all investigations of structural buckling in the plastic range. An extensive listing of plastic bifurcation analyses, including the initial discussion of reduced modulus versus tangent modulus load, are given in a survey paper by Sewell.[7]

A full understanding of bifurcation behaviour also requires an analysis of the behaviour after bifurcation, such as that given by Koiter[1] for elastic structures. In the plastic range such analyses are considerably complicated by the necessity to account for elastic unloading regions that start at bifurcation and subsequently spread into the material. An asymptotic theory of the initial post-bifurcation behaviour of structures in the plastic range has been developed by Hutchinson.[8,9] In this asymptotic theory details of the non-linear constitutive relationship play an important role, and the expansions developed by Hutchinson refer to a specific class of plasticity theories characterised by a smooth yield surface. For a material that develops a vertex on subsequent yield surfaces the asymptotic theory would have to be modified. This has not yet been done, but numerical computations of the effect of vertex formation on post-buckling behaviour have been presented recently by the authors.[10]

A natural extension of Hutchinson's initial post-bifurcation theory would be to include the effect of small initial imperfections and try to obtain asymptotic estimates of the imperfection-sensitivity. However, no simple asymptotic relation has yet been developed,[9,11] mainly due to the dominating effect of unloading on behaviour near the bifurcation load. On the other hand, if the approximating assumption is made that elastic unloading can be neglected, while all other non-linearities are still accounted for, asymptotic formulas can be obtained analogous with those known from elastic buckling theory. This approach has been used by Needleman and Tvergaard[12] for a rectangular plate and by Tvergaard[13] for a shell structure, and recently[10] the analysis has been extended to include the possibility of a geometrically non-linear pre-bifurcation state. In the present chapter the asymptotic analysis neglecting elastic unloading will be applied to estimate the imperfection-sensitivity of elastic–plastic columns.

Due to the highly non-linear nature of plastic buckling problems, numerical solutions have played an important role in elucidating plastic post-buckling behaviour. Although no attempt is made to review the growing literature in this field, some studies, primarily numerical, that illustrate aspects of plastic post-buckling behaviour are discussed. One feature which is frequently observed in plastic buckling experiments is that the final buckled configuration exhibits a localised buckling pattern in contrast to the periodic pattern associated with the bifurcation mode, e.g. Moxham.[14] The treatment of this phenomenon, recently given by the authors,[15] is briefly reviewed.

2. ANALYSIS OF BIFURCATION AND POST-BIFURCATION

A full description of structural buckling behaviour requires consideration of post-buckling phenomena in addition to a determination of the critical load. For elastic structures, subject to conservative loading, a thorough understanding of post-bifurcation behaviour has been achieved based on Koiter's general theory of elastic stability (see reference 1 and also Budiansky, reference 16). When the loading is characterised by a single load parameter, λ, and when bifurcation is associated with a unique mode, the initial post-bifurcation behaviour is described by an asymptotically exact expression of the form

$$\lambda = \lambda_c + \lambda_1^e \xi + \lambda_2^e \xi^2 + \cdots \tag{1}$$

Here, λ_c is the bifurcation load and ξ the bifurcation mode amplitude. The parameters λ_1^e and λ_2^e characterise the initial post-bifurcation behaviour and geometrical non-linearities play a crucial role in determining their values. Within the framework of the general theory of elastic stability, the behaviour of a slightly imperfect structure can be related to the post-bifurcation behaviour of the corresponding perfect structure. If $\lambda_1^e \neq 0$, or

$\lambda_1^e = 0$ with $\lambda_2^e < 0$, the structure is imperfection-sensitive in that the maximum support load, λ_s, of a slightly imperfect realisation of the structure can be significantly less than λ_c. Denoting the amplitude of the initial imperfection in the shape of the critical bifurcation mode by $\bar{\xi}$, and restricting attention to the special case of a linear pre-bifurcation state, the asymptotic expression for λ_s is given by[1,16]

$$\lambda_s/\lambda_c = 1 - 2 \left| \frac{\lambda_1^e}{\lambda_c} \right|^{1/2} |\bar{\xi}|^{1/2} \qquad \text{for } \lambda_1^e \bar{\xi} < 0$$

$$\lambda_s/\lambda_c = 1 - 3 \left| \frac{\lambda_2^e}{4\lambda_c} \right|^{1/3} \bar{\xi}^{2/3} \qquad \text{for } \lambda_1^e = 0, \ \lambda_2^e < 0 \qquad (2)$$

For structures which can support stresses beyond the yield stress a similarly comprehensive framework for analysing post-buckling behaviour and imperfection-sensitivity is not yet available. A major impediment to such a development stems from the non-linear, although homogeneous of degree one, dependence of the stress rate on the strain rate that characterises an elastic–plastic solid. Indeed, this non-linear dependence leads to the fundamental distinction, in the plastic range, between loss of uniqueness and loss of stability elucidated by Shanley.[4] A thorough discussion of plastic column buckling including a detailed development of Shanley's[4] justification of the tangent modulus load as the lowest bifurcation load of a column compressed into the plastic range has been given by Sewell.[7] Hill's[5,6] theory of uniqueness and bifurcation in the plastic range generalises Shanley's[4] analysis to three dimensional continua. Reviews of Hill's theory are included in the papers of Sewell[7] and Hutchinson.[9]

Here, the application of Hill's[5,6] theory in analyses of the plastic buckling of thin-walled plate and shell structures will be outlined within the framework of the Donnell-Mushtari-Vlasov (DMV) theory of plates and shells. The prescribed loads and the prescribed displacements, if any, are taken to be proportional to a single monotonically increasing parameter, λ. With A denoting the area of the shell middle surface, the incremental principle of virtual work can be written as

$$\int_A [\dot{N}^{\alpha\beta}\delta\dot{\varepsilon}_{\alpha\beta} + \dot{M}^{\alpha\beta}\delta\dot{\kappa}_{\alpha\beta} + N^{\alpha\beta}\dot{w}_{,\alpha}\delta\dot{w}_{,\beta}] \, dA = \dot{\lambda}(EVW) \qquad (3)$$

Here, and subsequently, Greek indices range from 1 to 2, the subscript ',α' denotes covariant differentiation with respect to the coordinates x^α on the middle surface and a superposed dot, denotes differentiation with respect to some monotonically increasing parameter that characterises the loading history. In eqn (3), EVW represents the external virtual work due to prescribed distributions of dead surface loads and boundary loads, $N^{\alpha\beta}$ are the membrane stresses, $M^{\alpha\beta}$ are the bending moments and the expressions for the strain increments $\dot{\varepsilon}_{\alpha\beta}$ and $\dot{\kappa}_{\alpha\beta}$ in DMV theory are

$$\dot{\varepsilon}_{\alpha\beta} = \tfrac{1}{2}(\dot{u}_{\alpha,\beta} + \dot{u}_{\beta,\alpha}) - d_{\alpha\beta}\dot{w} + \tfrac{1}{2}w_{,\alpha}\dot{w}_{,\beta} + \tfrac{1}{2}\dot{w}_{,\alpha}w_{,\beta} \tag{4}$$

$$\dot{\kappa}_{\alpha\beta} = \dot{w}_{,\alpha\beta} \tag{5}$$

where u_α and w are the inplane and normal displacements, respectively, and $d_{\alpha\beta}$ are the covariant components of the curvature tensor for the undeformed middle surface.

One possible equilibrium state, termed the fundamental state, is assumed to be available for all values of λ and to be unique for sufficiently small λ. Field quantities associated with this state are denoted by a zero sub- or superscript, e.g. $N_0^{\alpha\beta}$, w^0, etc.

At some given stage of the loading history the fundamental state is taken to be known and a load increment $\dot{\lambda}$ is prescribed. Suppose that two incremental equilibrium states are possible for this prescribed load increment. From the fact that each possible equilibrium state satisfies eqn (3) and satisfies the same displacement boundary conditions, it can be shown that the following identity holds[5,6,9]

$$H = \int_A (\tilde{N}^{\alpha\beta}\delta\tilde{\varepsilon}_{\alpha\beta} + \tilde{M}^{\alpha\beta}\delta\tilde{\kappa}_{\alpha\beta} + N_0^{\alpha\beta}\tilde{w}_{,\alpha}\delta\tilde{w}_{,\beta})\,\mathrm{d}A = 0 \tag{6}$$

where the tilde denotes the difference between field quantities associated with two possible solutions and $\tilde{\varepsilon}_{\alpha\beta}$ and $\tilde{\kappa}_{\alpha\beta}$ are given by eqn (4) and eqn (5) with the superposed dot replaced by a tilde and with $w = w^0$.

The stress state in the shell is approximately plane so that only the inplane stresses enter into the stress–strain relation. Thus, for an elastic–plastic solid,

$$\dot{\sigma}^{\alpha\beta} = L^{\alpha\beta\gamma\delta}\dot{\eta}_{\gamma\delta} \tag{7}$$

where $\dot{\eta}_{\gamma\delta} = \dot{\varepsilon}_{\gamma\delta} - x^3\dot{\kappa}_{\gamma\delta}$ is the strain increment at distance x^3 from the middle surface. The instantaneous moduli, $L^{\alpha\beta\gamma\delta}$, are homogeneous of degree zero in the stress increments and are assumed to have the symmetries $L^{\alpha\beta\gamma\delta} = L^{\gamma\delta\alpha\beta} = L^{\alpha\beta\delta\gamma}$. When the current stress state lies on the yield surface, the instantaneous moduli for an elastic–plastic solid with a smooth yield surface, such as the widely employed J_2 flow theory, have two branches; one branch corresponding to plastic loading and the other to elastic unloading. In this case the plane stress instantaneous moduli $L^{\alpha\beta\gamma\delta}$ can be written in the form (Hutchinson[9])

$$L^{\alpha\beta\gamma\delta} = L_e^{\alpha\beta\gamma\delta} - \alpha g^{-1} m^{\alpha\beta} m^{\gamma\delta} \qquad (8)$$

where $L_e^{\alpha\beta\gamma\delta}$ are the plane stress elastic moduli. If the current stress state is on the yield surface and if $m^{\alpha\beta}\dot{\eta}_{\alpha\beta} \geq 0$ then $\alpha = 1$, otherwise $\alpha = 0$. A fundamental role is played in Hill's[5,6] theory by the concept of a linear comparison solid. At the current load level, the linear comparison solid has instantaneous moduli $L_c^{\alpha\beta\gamma\delta}$ equal to the plastic moduli in the current plastic zone, regardless of the sign of $m^{\alpha\beta}\dot{\eta}_{\alpha\beta}$, and equal to the elastic moduli elsewhere. Since the instantaneous moduli of the linear comparison solid are independent of the stress rate, the functional governing bifurcation for the linear comparison solid is quadratic. When the fundamental state corresponds to a membrane state of stress, in a shell with thickness h, this functional takes the form

$$F = \int_A \left[hL_c^{\alpha\beta\gamma\delta}\tilde{\varepsilon}_{\alpha\beta}\tilde{\varepsilon}_{\gamma\delta} + \frac{h^3}{12} L_c^{\alpha\beta\gamma\delta}\tilde{\kappa}_{\alpha\beta}\tilde{\kappa}_{\gamma\delta} + N_0^{\alpha\beta}\tilde{w}_{,\alpha}\tilde{w}_{,\beta} \right] dA \qquad (9)$$

Presuming that the prescribed loading increment gives rise to plastic loading everywhere in the current plastic zone and that g in eqn (8) is positive, corresponding to material strain hardening, the integrand of H is never smaller than that of F (see Hill[5,6] and Hutchinson[9]). Hence,

$$H \geq F \qquad (10)$$

Consequently, if $F > 0$ for all admissible non-vanishing displacement fields, the solution for the elastic–plastic solid is unique.

The lowest eigenvalue, λ_c, of F gives the bifurcation load for the linear comparison solid. The lowest possible bifurcation load for the elastic–plastic structure is at λ_c if H (eqn (6)), vanishes simultaneously with F (eqn (9)), and this, almost always, implies bifurcation with increasing load, in accord with Shanley's[4] concept.[5,6,9]

The initial post-bifurcation behaviour around λ_c is given by

$$\lambda = \lambda_c + \lambda_1\xi + \cdots \qquad (11)$$

where $\xi\,(>0)$ is the bifurcation mode amplitude, and must be chosen large enough to ensure that the linear combination of the fundamental solution increment and ξ times the bifurcation mode gives plastic loading throughout the current plastic zone. In fact, Hutchinson[8,9] has shown that λ_1 must usually be identified as the smallest value consistent with this plastic loading constraint.

Hutchinson[8,9] has also developed a boundary layer analysis which follows the propagation of the unloading zone into the material and this analysis extends the expansion, eqn (11), to

$$\lambda = \lambda_c + \lambda_1 \xi + \lambda_2 \xi^{1+\beta} + \cdots \qquad (12)$$

The coefficient λ_2 in eqn (12) is necessarily negative and the value of β depends on the shape of the unloading zone propagating into the material. In the most common case, unloading starts at an isolated point on a smooth surface and $\beta = \frac{1}{3}$. In any case, $0 < \beta < 1$, and from eqn (12), it follows that although $d\lambda/d\xi = \lambda_1 > 0$ at bifurcation, the curvature $d^2\lambda/d\xi^2$ is negative and is infinite initially. Thus, bifurcation takes place under increasing load, but Hutchinson's[8,9] analysis indicates that a maximum load is attained at finite, although generally small, buckling deflections.

Our discussion so far has focused on elastic–plastic solids with a smooth yield surface. However, in the late 1940s and early 1950s it was noted that calculations employing the simplest smooth yield surface flow theory (J_2 flow theory) could give bifurcation loads greatly in excess of experimental buckling loads. On the other hand, calculations based on the corresponding deformation theory, which is not a physically acceptable plasticity theory, gave much better agreement with test results, leading to what has been termed the 'plastic buckling paradox'. A comprehensive discussion of this controversy and the issues raised is given by Hutchinson.[9]

The cruciform column, which played a central role in this controversy, is probably the most extreme example of the discrepancy between flow theory and deformation theory bifurcation predictions. For a sufficiently long cruciform column, the bifurcation mode is a pure twisting mode and for any isotropic smooth yield surface plasticity theory, the critical stress is given by

$$\sigma_c = G \left(\frac{h}{b}\right)^2 \qquad (13)$$

where G is the elastic shear modulus, h is the plate thickness and b is the plate width (see Fig. 1). For an isotropic deformation theory the critical stress is given by an expression of the form of eqn (13) with the elastic shear modulus replaced by the modulus

$$G_s = \frac{G}{1 + 3G\left(\dfrac{1}{E_s} - \dfrac{1}{E}\right)} \qquad (14)$$

where E_s is the current secant modulus (see reference 9). Experimentally

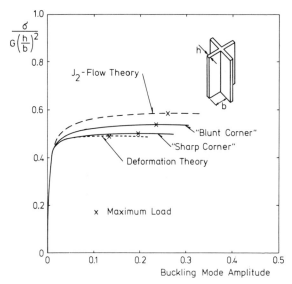

FIG. 1.   Influence of a small imperfection on a cruciform column for various
constitutive models.

obtained buckling loads for cruciform columns are in good agreement with
the predictions of deformation theory, but not with those of the simplest
flow theory. Attempts to resolve this discrepancy have taken essentially two
lines of attack.

Onat and Drucker[17] argued that unavoidably small initial imperfections
could bring the maximum support load given by a smooth yield surface
theory in line with the experimentally observed values. More recently,
Hutchinson and Budiansky[18] have presented a detailed analysis of the
imperfection sensitivity of the cruciform column, employing J_2 flow
theory, and conclude that for moderately or lightly strain hardening
materials, unavoidably small initial imperfections would reduce the
maximum support load to the deformation theory value, but that for a
relatively highly hardening material the imperfection magnitudes required
to do so are not necessarily unavoidable.

The other approach to resolving this discrepancy, initiated by Batdorf,[19]
pursued justification of the deformation theory calculations in terms of a
sophisticated flow theory that permits the development of a vertex on the
yield surface. The occurrence of vertices on yield surfaces is supported by
models of the elastic–plastic behaviour of polycrystalline metals based on
single crystal slip (Hill,[20] Hutchinson,[21] Lin[22]). However, experimental

evidence concerning vertex formation on subsequent yield surfaces is ambiguous and conflicting (Michno and Findley,[23] Hecker[24]).

Incremental theories based on a system of linear loading functions have been constructed which give bifurcation predictions, from nearly proportional loading paths, identical to the simplest deformation theory (Sanders[25] and Sewell[26]). More recently a phenomenological corner theory of plasticity, termed J_2 corner theory, has been developed by Christoffersen and Hutchinson.[27] In this theory the instantaneous moduli for nearly proportional loading are chosen equal to the J_2 deformation theory moduli and for increasing deviation from proportional loading the moduli 'stiffen' monotonically until they coincide with the elastic moduli for stress increments directed along or within the corner of the yield surface. This corner theory was introduced to provide a phenomenological theory that embodied features of physical theories, while being tractable enough to employ in analyses of post-bifurcation and imperfection-sensitivity.

Hill's[5,6] theory of uniqueness and bifurcation for elastic–plastic solids is broad enough to encompass a wide class of plasticity theories with cornered yield surfaces, including the theories mentioned above.[25–27] In essence, these corner theories are constructed so that J_2 deformation theory is the appropriate linear comparison solid when bifurcation takes place from nearly proportional loading paths.

Needleman and Tvergaard[10] have studied the imperfection-sensitivity of the cruciform column using the J_2 corner theory of Christofferson and Hutchinson.[27] Figure 1 depicts their results for a case when the ratio of the corner theory bifurcation stress to the smooth yield surface flow theory bifurcation stress (eqn (13)) is $\frac{1}{2}$. The analysis is based on DMV theory and also employs further kinematic approximations used by Hutchinson and Budiansky.[18] The details of the corner descriptions employed in reference 10 will not be given here, but we note that each case considered in Fig. 1 has an identical uniaxial stress–strain curve as well as an identical initial imperfection amplitude. For the 'sharp' vertex, the maximum load is only slightly higher than that given by deformation theory, while for the more blunt' vertex the maximum load is about halfway between the deformation theory and flow theory values. For all cases considered in reference 10, the corner theory predictions fall between the J_2 deformation theory predictions and the J_2 flow theory predictions. In situations such as the one discussed above, where the bifurcation predictions of simplest smooth yield surface flow theory and deformation theory differ substantially, the post-buckling behaviour and imperfection-sensitivity can be expected to be sensitive to the details of the constitutive description.

In such circumstances analyses based on a smooth yield surface plasticity theory are expected to exhibit an extreme sensitivity to small initial imperfections, arising from the choice of constitutive law. This type of imperfection-sensitivity, associated with the 'plastic buckling paradox' has received considerable attention. When a smooth yield surface theory is used in analysing problems in this class, the maximum support load with a very small imperfection may be so far below the bifurcation load that the bifurcation load is of no significance. However, when a vertex theory is employed the bifurcation load retains significance.

It should be emphasised that imperfection-sensitivity is also found in the plastic range in cases where the bifurcation predictions of simplest flow theory and deformation theory do not differ substantially, even for structures such as flat plates which are imperfection insensitive in the elastic range.

3.   ASYMPTOTIC ANALYSIS NEGLECTING ELASTIC UNLOADING

The use of a hypoelastic asymptotic analysis to estimate the imperfection-sensitivity of an elastic–plastic structure was first introduced by Hutchinson and Budiansky,[18] for the case of a cruciform column. This structure is exceptional in that often no strain rate reversal occurs before the maximum load, so that neglecting elastic unloading, and thus replacing the elastic–plastic material model by a corresponding hypoelastic material model, is completely justified. For most other structures compressed into the plastic range elastic unloading does occur before the load maximum, and this results in considerable extra complications that have so far impeded the attainment of a simple asymptotic estimate of imperfection-sensitivity.[9,11]

The possibility of also using the hypoelastic asymptotic analysis in cases where neglecting elastic unloading is not rigorously justified has been investigated by Needleman and Tvergaard.[12] It was found, for an axially compressed square plate and subsequently also for a cylindrical panel,[13] that the hypoelastic asymptotic estimates can reveal the 'overall' behaviour when small initial imperfections are present. The analysis has recently been extended to include the possibility of a geometrically non-linear pre-bifurcation state and to cover cases of asymmetric buckling.[10]

Another approximate approach that accounts for elastic unloading by using an asymptotic estimate of the locus of maxima around the reduced modulus load has been suggested by van der Heijden.[28] This method gives a useful representation of the imperfection-sensitivity in the case of a model problem, but has not yet been generalised to more realistic structures.

A few of the main expressions resulting from the hypoelastic asymptotic analysis shall be stated here. In cases where the buckling mode is unique, the initial post-bifurcation behaviour of the perfect structure is specified by an asymptotic expression for the load parameter λ in terms of the buckling mode amplitude ξ analogous to eqn (1)

$$\lambda = \lambda_c + \lambda_1^{he}\xi + \lambda_2^{he}\xi^2 + \cdots \tag{15}$$

Here, λ_c is the buckling load (identical with that of the elastic–plastic structure), and the parameters λ_1^{he} and λ_2^{he} characterise the post-bifurcation behaviour.[10,12,13] If the structure has an initial geometrical imperfection, shaped as the critical bifurcation mode with amplitude $\bar{\xi}$, snap buckling may occur at a load λ_s smaller than λ_c. For a structure with asymmetric bifurcation ($\lambda_1^{he} \neq 0$) the following asymptotic estimate of λ_s is found, corresponding to $\lambda_1^{he}\bar{\xi} < 0$

$$\left.\begin{aligned}
\frac{\lambda_s}{\lambda_c} &= 1 - \mu|\bar{\xi}|^{1/(\psi+1)} \\
\mu &= (\psi+1)\psi^{-\psi/(\psi+1)}\left[\frac{|\lambda_1^{he}|}{\lambda_c}\right]^{1/(\psi+1)}[p(\lambda_c)]^{1/(\psi+1)}
\end{aligned}\right\} \tag{16}$$

whereas no λ_s is predicted, to lowest order, for $\lambda_1^{he}\bar{\xi} > 0$. If the post-bifurcation behaviour is symmetric ($\lambda_1^{he} = 0$) imperfection-sensitivity is predicted for $\lambda_2^{he} \geq 0$, but the following estimate is obtained for $\lambda_2^{he} < 0$ and $\lambda_1^{he} = 0$

$$\left.\begin{aligned}
\frac{\lambda_s}{\lambda_c} &= 1 - \mu\bar{\xi}^{2/(2\psi+1)} \\
\mu &= (2\psi+1)(2\psi)^{-2\psi/(2\psi+1)}\left[\frac{-\lambda_2^{he}}{\lambda_c}\right]^{1/(2\psi+1)}[p(\lambda_c)]^{2/(2\psi+1)}
\end{aligned}\right\} \tag{17}$$

In the expressions (16) and (17) the value $p(\lambda_c)$ is to be determined from a regular perturbation analysis. For an elastic (linear or non-linear) solid, $\psi = 1$, and expressions (16) and (17) reduce to their elastic counterparts, expressions (2), while for a path dependent hypoelastic solid $\psi > 1$.

Here, the asymptotic analysis will be illustrated by investigating the imperfection-sensitivity of plastic columns. This case is considerably less complex than the plate and shell problems treated previously;[12,13] but it does incorporate the basic approximation in the hypoelastic analysis of including all geometric and material non-linearities except for elastic unloading. Furthermore, in the case of a simple column, where only a uniaxial stress state is accounted for, the hypoelastic material reduces to a non-linear elastic material. This means that the analysis to be presented is also covered by Koiter's elastic post-buckling theory,[1,16] although this theory has mainly been applied to linearly elastic problems.

The columns considered have a length l and are simply supported at the ends. A Cartesian coordinate system is chosen so that the column centre line is the x_1-axis and the cross-sections are symmetric about the x_2-axis. The cross-sections are taken to have the height h and the varying width $b(x_2)$, as shown in Fig. 2. The cross-sectional area is denoted A, the area moment of inertia is denoted I and the radius of gyration of the cross-section is $r = \sqrt{I/A}$.

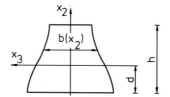

FIG. 2.   Column cross-section.

With the displacements of the centre line denoted by u and w in the x_1 and x_2 directions, respectively, the strain ε and the bending strain κ of the centre line are taken to be

$$\varepsilon = u_{,1} + \tfrac{1}{2}w_{,1}w_{,1} \qquad \kappa = w_{,11} \tag{18}$$

The beam approximation to the axial strain at coordinate x_2 is

$$\eta = \varepsilon - x_2\kappa \tag{19}$$

For a material with elastic unloading neglected, the incremental relationship between the normal stress σ and the strain η is

$$\dot{\sigma} = E_t\dot{\eta} \tag{20}$$

where E_t is the tangent modulus at the current stress level, and the superposed dot denotes differentiation with respect to a loading parameter. The resultant force and the bending moment corresponding to the stresses σ on a cross-section are denoted by N and M, respectively.

The uniform stress state in a perfect column prior to bifurcation is $\sigma_0 = (\lambda/\lambda_c)\sigma_c$, where the '$c$' subscript denotes the value at the critical bifurcation point. The critical stress σ_c and the corresponding bifurcation mode $\overset{(1)}{w}$ are given by

$$\sigma_c = -\pi^2 E_t^c \left(\frac{r}{l}\right)^2 \qquad \overset{(1)}{w} = r\sin\frac{\pi x_1}{l} \qquad (21)$$

An asymmetric post-bifurcation behaviour has been found by Tvergaard and Needleman,[29] both asymptotically and numerically, for elastic–plastic columns with asymmetric cross-sections. The corresponding values of the parameter λ_1^{he} were found to be given by the expression

$$\lambda_1^{he}/\lambda_c = \frac{4\rho}{3\pi}\frac{q}{1+q} \qquad (22)$$

where

$$\rho = \frac{1}{rI}\int_{-d}^{h-d} x_2^3 b(x_2)\,\mathrm{d}x_2 \qquad q = \pi^2 \left(\frac{r}{l}\right)^2 \frac{\mathrm{d}E_t}{\mathrm{d}\sigma}\bigg|_c \qquad (23)$$

Thus, in order to use expressions (16), ψ and $p(\lambda_c)$ remain to be determined, and since the material considered is purely elastic, $\psi = 1$ is known *a priori*. An analysis leading to the last parameter, $p(\lambda_c)$, is shown in the following paragraphs.

The effect of a small imperfection, $\bar{w} = \bar{\xi}\overset{(1)}{w}$, at the initial stage of loading, for $\lambda < \lambda_c$, is found by a regular perturbation analysis. The displacements, strains and stresses are written as

$$u = u_0 + \tilde{u} \qquad w = w_0 + \tilde{w} \qquad \eta = \eta_0 + \tilde{\eta} \qquad \sigma = \sigma_0 + \tilde{\sigma} \qquad (24)$$

where the solution for the perfect structure at the current load level λ is denoted by subscript zero, while a tilde denotes small perturbation quantities that vanish at $\lambda = 0$. All equations are linearised with respect to the small perturbation and imperfection quantities. Thus, using the solution for the perfect structure to eliminate the external virtual work, the linearised principle of virtual work takes the form

$$\int_0^l \{\tilde{N}\delta e + \tilde{M}\delta\kappa + N_0(\tilde{w}_{,1} + \bar{\xi}\overset{(1)}{w}_{,1})\delta w_{,1}\}\,\mathrm{d}x_1 = 0 \qquad (25)$$

where e is the linear part of the strain, ε. The linearised relationship between perturbation stresses and strains for the general hypoelastic solid is obtained from an incremental constitutive law and requires the solution of differential equations. However, for the present elastic solid, the known total form of the stress–strain relationship yields directly the linearised expression

$$\tilde{\sigma} = E_t^0 \tilde{\eta} \tag{26}$$

Axial equilibrium for the perfect column and for the imperfect column, at the current value of λ, require that both N_0 and $N_0 + \tilde{N}$ are equal to $(\lambda/\lambda_c)N_c$, and thus that $\tilde{N} \equiv 0$. The remaining part of eqn (25) is solved by using eqn (26) to evaluate $\tilde{M}$ and by expressing $\tilde{w}$ in terms of an infinite series

$$\tilde{w} = \sum_{n=1}^{\infty} \xi_n r \sin \frac{n\pi x_1}{l} \tag{27}$$

When choosing $\delta w = \overset{(1)}{w}$, and thus $\delta \kappa = \overset{(1)}{\kappa}$, in eqn (25), all terms except the first in this series vanish by the extended orthogonality that was also used in references 12 and 13, but which was not assumed in reference 10. Thus, the growth of the bifurcation mode amplitude $\xi(\lambda) = \xi_1$ is obtained from

$$\int_0^l \left\{ E_t^0 I \overset{(1)(1)}{\kappa \kappa} + \frac{\lambda}{\lambda_c} N_c(\xi + \bar{\xi}) \overset{(1)}{w}_{,1} \overset{(1)}{w}_{,1} \right\} dx_1 = 0 \tag{28}$$

In the general hypoelastic analysis we find that the initial growth of the mode amplitude can be written in the form

$$\xi(\lambda) = \bar{\xi} \left(1 - \frac{\lambda}{\lambda_c} \right)^{-\psi} p(\lambda) \tag{29}$$

where $p(\lambda)$ is finite at $\lambda = \lambda_c$. Substitution of this into eqn (28) gives

$$p(\lambda) = \frac{\dfrac{\lambda}{\lambda_c} \left(1 - \dfrac{\lambda}{\lambda_c} \right)}{\dfrac{E_t^0}{E_t^c} - \dfrac{\lambda}{\lambda_c}} \tag{30}$$

Finding the value $p(\lambda_c)$ to be used in expressions (16) and (17) requires knowledge of the λ − dependence of E_t^0. Here, the stress–strain curve for uniaxial tension is taken to be given by the power hardening law, also used in reference 29

$$\varepsilon = \frac{\sigma_e}{E} \qquad \text{for } \sigma_e \leq \sigma_y$$

$$\varepsilon = \frac{\sigma_y}{E}\left[\frac{1}{n}\left(\frac{\sigma_e}{\sigma_y}\right)^n - \frac{1}{n} + 1\right] \qquad \text{for } \sigma_e > \sigma_y$$

(31)

where E is Young's modulus, σ_y is the initial yield stress, n is the strain hardening exponent, and $\sigma_e = |\sigma|$. Application of eqn (31) in eqn (30) gives

$$p(\lambda_c) = 1/n \tag{32}$$

Numerical results for elastic–plastic columns with U-shaped or triangular cross-sections are given in reference 29, where details of the column geometries can be found. In Fig. 3 these results are compared with asymptotic predictions obtained from expressions (16). The column in Fig. 3(a) with a U-shaped cross-section is characterised by the parameters $-\sigma_c/\sigma_y = 1\cdot07$, $\lambda_1^{he}/\lambda_c = 0\cdot370$, $n = 10$, and thus $\mu = 0\cdot385$. The same U-shaped cross-section appears in Fig. 3(b), with $-\sigma_c/\sigma_y = 1\cdot19$, $\lambda_1^{he}/\lambda_c = 0\cdot309$, $n = 4$, $\mu = 0\cdot556$, and Fig. 3(c) represents a column with triangular cross-section and parameters $-\sigma_c/\sigma_y = 1\cdot07$, $\lambda_1^{he}/\lambda_c = 0\cdot216$, $n = 10$, $\mu = 0\cdot294$. Particularly in Fig. 3(b) the asymptotic predictions based on neglecting elastic unloading are in good agreement with the numerical results for the elastic–plastic column. For the more lightly hardening materials in Figs 3(a) and (c), where buckling of the perfect column occurs closer to the yield stress, the initial tendencies are well represented, but clearly the range of validity of the asymptotic predictions is smaller here.

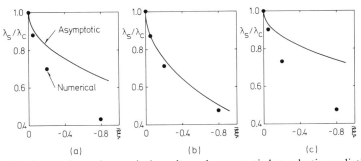

FIG. 3. Comparison of numerical results and asymptotic hypoelastic predictions for the imperfection-sensitivity of elastic–plastic columns. (a) U-shaped cross-section, $n = 10$ and $-\sigma_c/\sigma_y = 1\cdot07$. (b) U-shaped cross-section, $n = 4$ and $-\sigma_c/\sigma_y = 1\cdot19$. (c) Triangular cross-section, $n = 10$ and $-\sigma_c/\sigma_y = 1\cdot07$.

Since the numerically determined post-bifurcation behaviour (see reference 29) drops below that predicted by eqn (15), the overestimate of the load carrying capacities for imperfect columns depicted in Fig. 3 is to be expected.

As a different example we shall investigate a case with asymmetric post-bifurcation behaviour. Then, with $\lambda_1^{he} = 0$, the lowest order contribution to the post-bifurcation behaviour eqn (15) depends on the parameter λ_2^{he}, with ψ still unity and $p(\lambda_c)$ given by eqn (32). When $\lambda_2^{he} < 0$, a sensitivity to small imperfections is predicted by the asymptotic expressions (17).

The asymptotic expansions of the displacements, corresponding to eqn (15), are of the form

$$
\left.
\begin{aligned}
w = w_0 + \xi \overset{(1)}{w} + \xi^2 \overset{(2)}{w} + \cdots \\[2mm]
u = u_0 + \xi \overset{(1)}{u} + \xi^2 \overset{(2)}{u} + \cdots
\end{aligned}
\right\}
\tag{33}
$$

Similar expansions are used for η, σ, N, M, etc., and the relations between stress and strain measures are found to be

$$
\left.
\begin{aligned}
\left.\frac{d\sigma_0}{d\lambda}\right|_c = E_t^c \left.\frac{d\eta_0}{d\lambda}\right|_c \\[2mm]
\overset{(1)}{\sigma} = E_t^c \overset{(1)}{\eta} \qquad \overset{(2)}{\sigma} = E_t^c \overset{(2)}{\eta} + \tfrac{1}{2}\overset{(1)}{\sigma}\left.\frac{dE_t}{d\sigma}\right|_c \overset{(1)}{\eta}
\end{aligned}
\right\}
\tag{34}
$$

where $\lambda_1^{he} = 0$ is used here and in the following. In terms of the quantities defined by these expressions the coefficient λ_2^{he} is given by[10,12,13]

$$
\lambda_2^{he} = -\mathscr{C}/\mathscr{D}
\tag{35}
$$

$$
\mathscr{C} = \int_0^l \left\{ 3\overset{(2)}{N}\overset{(1)}{w}_{,1}\overset{(1)}{w}_{,1} + 6\overset{(1)}{N}\overset{(1)}{w}_{,1}\overset{(2)}{w}_{,1} \right\} dx_1
$$

$$
+ \int_0^l \int_{-d}^{h-d} \left\{ \overset{(2)}{\sigma}\left.\frac{dE_t}{d\sigma}\right|_c \overset{(1)}{\eta}\overset{(1)}{\eta} + 2\overset{(1)}{\sigma}\left.\frac{dE_t}{d\sigma}\right|_c \overset{(1)}{\eta}\overset{(2)}{\eta} + \tfrac{1}{2}\overset{(1)}{\sigma}\overset{(1)}{\sigma}\left.\frac{d^2E_t}{d\sigma^2}\right|_c \overset{(1)}{\eta}\overset{(1)}{\eta} \right\}
$$

$$
\times b(x_2)\,dx_2\,dx_1
\tag{36}
$$

$$
\mathscr{D} = \int_0^l \left\{ 6\overset{(1)}{N}\left.\frac{dw_{0,1}}{d\lambda}\right|_c \overset{(1)}{w}_{,1} + 3\left.\frac{dN_0}{d\lambda}\right|_c \overset{(1)}{w}_{,1}\overset{(1)}{w}_{,1} \right\} dx_1
$$

$$
+ \int_0^l \int_{-d}^{h-d} \left\{ \left.\frac{d\sigma_0}{d\lambda}\right|_c \left.\frac{dE_t}{d\sigma}\right|_c \overset{(1)}{\eta}\overset{(1)}{\eta} + 2\overset{(1)}{\sigma}\left.\frac{dE_t}{d\sigma}\right|_c \left.\frac{d\eta_0}{d\lambda}\right|_c \overset{(1)}{\eta} \right\}
$$

$$
\times b(x_2)\,dx_2\,dx_1
\tag{37}
$$

To obtain the value of λ_2^{he} from eqns (35)–(37) the second order perturbation contributions must be determined. From the axial equilibrium requirement, $N = N_0 = (\lambda/\lambda_c)N_c$, we directly find $\overset{(1)}{N} \equiv \overset{(2)}{N} \equiv 0$, which determines $\overset{(1)}{\varepsilon}$ and $\overset{(2)}{\varepsilon}$ by integration of expressions (34) over the cross-section. Furthermore, $w_0 \equiv 0$ is known *a priori*, and the boundary value problem for the second order displacement fields gives $\overset{(2)}{w}$ proportional with $\overset{(1)}{w}$. Thus, by the requirement of orthogonality $\overset{(2)}{w} \equiv 0$ is found.

Now, all quantities appearing in eqns (36) and (37) are known, and the result is

$$\mathscr{C} = \frac{3lr^4}{16}\left(\frac{\pi}{l}\right)^8 E_t^c\left\{-3\left(\frac{\mathrm{d}E_t}{\mathrm{d}\sigma}\bigg|_c\right)^2 r^2 I + \left[\left(\frac{\mathrm{d}E_t}{\mathrm{d}\sigma}\bigg|_c\right)^2 + E_t^c\frac{\mathrm{d}^2 E_t}{\mathrm{d}\sigma^2}\bigg|_c\right]K\right\} \quad (38)$$

$$\mathscr{D} = \frac{3\pi^2 r^2}{2l\lambda_c}\left\{N_c + \left(\frac{\pi}{l}\right)^2 \sigma_c\frac{\mathrm{d}E_t}{\mathrm{d}\sigma}\bigg|_c I\right\} \quad (39)$$

where the cross-section constant K is given by

$$K = \int_{-d}^{h-d} x_2^4 b(x_2)\,\mathrm{d}x_2 \quad (40)$$

From eqns (22) and (23) it is seen that $\lambda_1^{he} = 0$ for all columns with cross-sections symmetric about the x_3-axis. Using the stress–strain law, expressions (31), the tangent modulus and its derivatives at $\sigma = -\sigma_e$ are $E_t = E(\sigma_e/\sigma_y)^{1-n}$, $\mathrm{d}E_t/\mathrm{d}\sigma = (E/\sigma_y)(n-1)(\sigma_e/\sigma_y)^{-n}$ and $\mathrm{d}^2 E_t/\mathrm{d}\sigma^2 = (E/\sigma_y^2)(n-1)n(\sigma_e/\sigma_y)^{-n-1}$. Substituting the above expressions into eqns (38) and (39), we obtain

$$\frac{\lambda_2^{he}}{\lambda_c} = \frac{n-1}{8n}\{\Gamma(2n-1) - 3(n-1)\} \quad (41)$$

where

$$\Gamma = \frac{K}{r^2 I} \quad (42)$$

For a rectangular cross-section, $K = bh^5/80$ and $r^2 I = bh^5/144$, giving $\Gamma = 1.8$, so that the initial post-bifurcation behaviour is characterised by $\lambda_2^{he} > 0$ for any $n \geq 1$. This surprising result is closely related to the assumed stress–strain law. According to eqn (41) a negative λ_2^{he} is only possible for a column with a smaller value of the cross-section constant Γ. An idealised sandwich column with all material contained in two identical flanges has $\Gamma = 1$ and thus $\lambda_2^{he} < 0$ for $n > 2$.

For the rectangular cross-section, with $\Gamma = 1\cdot8$, eqn (38) shows that the positive value of λ_2^{he} and thus of $\mathscr{C}$ is dependent on the positive $d^2 E_t/d\sigma^2$ for the particular stress–strain law, expressions (31), used here. For a material assumed to harden according to a Ramberg-Osgood law there are stress ranges, in which $d^2 E_t/d\sigma^2$ is negative, so that a negative λ_2^{he} would be found, also for a column with a rectangular cross-section. Furthermore, for a linear hardening material both λ_1^{he} and λ_2^{he} vanish, whatever the shape of the cross-section.

The result found for a rectangular cross-section and the power hardening law, expressions (31), shall be discussed in more detail here, by analysing a special case. The column considered is defined by $h/l = 0\cdot05$, $\sigma_y/E = 0\cdot001$ and $n = 4$. For this column bifurcation occurs at $-\sigma_c/\sigma_y = 1\cdot20$ with $\lambda_2^{he}/\lambda_c = 0\cdot338$. The same column is analysed numerically by a linear incremental method as described in reference 29. In Fig. 4 the numerically determined variation of the load parameter λ versus the bifurcation mode amplitude ξ is plotted, using solid curves for elastic–plastic columns and dashed curves for columns in which elastic unloading is neglected.

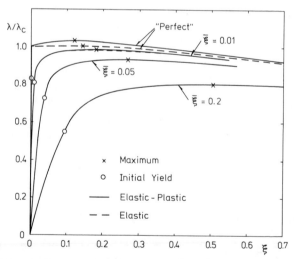

FIG. 4.   Load versus buckling mode displacement for elastic–plastic column with rectangular cross-section and $\sigma_y/E = 0\cdot001$, $n = 4$, $h/l = 0\cdot05$. Dashed curves show behaviour when elastic unloading is suppressed.

The numerically determined post-bifurcation behaviour for the non-linearly elastic column in Fig. 4 does initially show the stable behaviour predicted by the positive value of λ_2^{he}. There is even good quantitative agreement with the asymptotic prediction, eqn (15) up to $\xi \simeq 0.08$, where some of the material re-enters the linear part of the stress–strain curve, expressions (31); but subsequently a maximum is reached, and for larger mode amplitudes the post-bifurcation curve drops below the bifurcation load. With $\lambda_2^{he} > 0$ no imperfection induced reduction of the load carrying capacity is predicted asymptotically, to lowest order, and this holds true for very small imperfections. However, Fig. 4 shows that the column is sensitive to imperfection amplitudes larger than about 0.01, and thus the range of validity of the asymptotic expansion is quite limited in the present case. The fact that a Taylor expansion of the tangent modulus around the bifurcation point cannot account for the linear part of expressions (31) is an important part of the reason for the limited range of validity of the expansion, eqn (15).

The different shapes of the two post-bifurcation curves in Fig. 4 for small values of ξ clearly show the effect of elastic unloading. However, for an imperfection $\bar{\xi} = 0.01$ the curve obtained for the non-linearly elastic column is already very close to that obtained for the elastic–plastic column. For large imperfections the results for the non-linear elastic and the elastic–plastic columns are indistinguishable in Fig. 4. Quite similar results were obtained numerically in reference 29 for a column with a U-shaped cross-section. Furthermore, both in Fig. 4 and in reference 29 the post-bifurcation behaviour found for the non-linearly elastic (or hypoelastic) column is very close to that of the elastic–plastic column at values somewhat beyond the load maxima. Therefore it appears that, quantitatively, very little information about the elastic–plastic column buckling behaviour is lost by neglecting elastic unloading. In fact, the limitations due to this approximation seem less significant than the limited range of validity of an asymptotic expansion in such a highly non-linear problem.

4. DISCUSSION OF PLASTIC POST-BUCKLING BEHAVIOUR

Here we will discuss briefly the results of some studies, primarily numerical, which illustrate characteristic features of the post-buckling behaviour of plates and shells in circumstances where material plasticity plays a significant role. No attempt will be made to review the growing literature in this field. More complete references are given by Sewell,[7] Hutchinson,[9] and

Tvergaard.[3] In virtually all the examples to be discussed, the simplest smooth yield surface theory of plasticity (J_2 flow theory with isotropic hardening) was employed to characterise the material behaviour. However, attention here is focused on circumstances where the plastic pre-strains are small enough that the bifurcation predictions of simplest flow theory and simplest deformation theory do not differ significantly. Hence, the imperfection sensitivity exhibited in these cases arises primarily from the decreasing stiffness with increasing deformation characteristic of plasticity and is not the anomalous constitutive relation dependent imperfection sensitivity associated with the plastic buckling 'paradox' discussed in Section 2 in conjunction with the cruciform column.

For plate and shell structures, comparisons between the hypoelastic asymptotic analysis and the results of full numerical solutions have been carried out by Needleman and Tvergaard[12] and Tvergaard.[13] Due to the change in loading path associated with the bifurcation mode, material path dependence plays an essential role in the asymptotic analyses in references 12 and 13, even though elastic unloading is suppressed. This contrasts with the situation prevailing in the column problems discussed in the previous section, where neglect of elastic unloading leads to path independent (i.e. non-linearly elastic) material behaviour.

Needleman and Tvergaard[12] studied the post-buckling behaviour and imperfection sensitivity of square plates simply supported on all four sides and subject to axial compression in one direction. Two types of in-plane boundary conditions were considered. One set of in-plane boundary conditions required all four edges of the plate to remain straight while the other set left the edges unconstrained. In the elastic range the initial post-buckling behaviour can be determined using Koiter's[1] general theory and is stable symmetric for both types of in-plane boundary conditions, so that the load and buckling mode amplitude are related by eqn (1) with $\lambda_1^e = 0$ and $\lambda_2^e > 0$.

In the plastic range, the initial post-bifurcation behaviour, determined by means of Hutchinson's[8,9] asymptotic analysis, results in a load-buckling mode amplitude relation of the form of eqn (12), with λ_1 and λ_2 independent of the in-plane boundary conditions. With elastic unloading neglected, the initial post-bifurcation behaviour is given by eqn (15) with $\lambda_1^{he} = 0$. For a lightly hardening material ($n = 10$ in expressions (31)) λ_2^{he} is negative for both types of in-plane boundary conditions.[12] On the other hand, for a rather highly hardening material ($n = 3$ in expressions (31)) λ_2^{he} is positive when the plate edges are unconstrained and negative when the plate edges are unconstrained. The corresponding numerically determined load-buckling deflection curves, with full account taken of elastic unloading, exhibit imperfection sensitivity with the unconstrained edge boundary conditions and imperfection insensitivity with the constrained edge boundary conditions. For the lightly hardening material, $n = 10$, imperfection sensitivity is found numerically for both sets of boundary conditions. Furthermore, for small initial geometric imperfections in the shape of the critical bifurcation mode, there is reasonably good quantitative agreement between the maximum support loads given by the expressions (17) and the numerically computed maximum support loads.

An analogous study for an axially compressed elastic–plastic cylindrical panel of the type occurring in stiffened shells has been carried out by Tvergaard.[13] In reference 13 attention was focused on local buckling between the stiffeners and the torsional rigidity of the stiffeners was neglected. Koiter[30] analysed the post-buckling behaviour of panels that bifurcate in the elastic range and showed that sufficiently flat panels have stable symmetric post-bifurcation behaviour while more curved panels are imperfection sensitive. For panels that bifurcate in the plastic range, Tvergaard[13] finds good agreement between the maximum support loads predicted by the asymptotic analysis and those given by a full numerical analysis for a wide range of panel curvatures and material hardening rates, when the initial imperfection amplitude is small. However, there are combinations of material and geometric parameters, which for a perfect panel lead to an initial load maximum, a rather small load drop before a load minimum is attained and then an increasing load in the advanced post-buckling range. In such a case, no maximum support load is found for a panel with a small, but finite, initial imperfection. Another example of the effect of imperfections on the maximum support load, in which there is a complex post-bifurcation load deflection curve, is provided by Tvergaard's[31] numerical investigation of the imperfection sensitivity of elastic–plastic oval cylindrical shells under axial compression.

The plate and shell problems considered in references 12 and 13, in line with the behaviour exhibited in Fig. 4, indicate that relatively little quantitative information concerning initial post-buckling behaviour is lost by neglecting elastic unloading. However, due to the complex interaction between material and geometric non-linearities that occur in the plastic range, the behaviour of a structure with a small, but finite, imperfection is not necessarily simply related to the initial post-bifurcation behaviour of the corresponding perfect structure. As illustrated in Fig. 4, phenomena occurring at finite buckling deflections may play a significant role. The major limitation on the validity of the hypoelastic asymptotic analysis arises from its asymptotic nature rather than from errors induced by neglecting elastic unloading. Such limitations on the validity of estimates of imperfection sensitivity based on asymptotic estimates valid near the bifurcation point do occur in the elastic range, but are apt to be more prevalent in the plastic range.

A different, but related, limitation on initial post-buckling analyses occurs when the stress level at bifurcation is close to, but less than, that required for initial yielding. For columns and flat plates which exhibit stable post-bifurcation behaviour in the elastic range the onset of plastic yielding, due to the compressive bending stresses induced by buckling, can lead to a maximum support load being achieved at finite, although possibly small, buckling deflections, along with the related imperfection sensitivity. The effect of material plasticity on the maximum support load of columns has been investigated by many authors, as referenced in the survey paper by Sewell.[7] More recently, a Perry type first-yield formula for imperfect columns with strain-hardening material behaviour has been investigated by Calladine.[32] An illustrative example of the influence of material plasticity on the load carrying capacity of columns which bifurcate in the elastic range is shown in Fig. 5 (taken from reference 29). A U-shaped column is considered which exhibits asymmetric post-bifurcation behaviour in the plastic range. The geometric parameters and elastic properties are chosen so that the ratio of bifurcation stress to yield stress is 0·991. An initial imperfection in the shape of the eigenmode is employed, with amplitude $\bar{\xi}$,

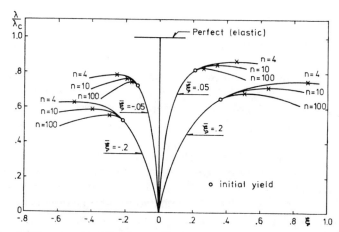

FIG. 5.   Load versus buckling mode displacement for columns with U-shaped cross-sections, $-\sigma_c/\sigma_y = 0.991$, and various values of the strain hardening exponent n.

and the load-buckling deflection curves for various strain-hardening exponents, n, in expressions (41) are determined. Even though the initial post-bifurcation behaviour in the elastic range is symmetric, the imperfection sensitivity is asymmetric due to plastic yielding. A Perry first-yield formula would also give asymmetric initial yielding (see Fig. 5). For the columns in Fig. 5 the maximum support load is only a little above initial yielding even for a material with fairly high strain-hardening, $n = 4$.

Even though flat plates generally exhibit much more stable post-bifurcation behaviour in the elastic range than columns, a significant reduction in load carrying capacity can occur due to plastic yielding. Figure 6, taken from reference 33, illustrates the effect of plastic yielding on load carrying capacity for a simply supported circular plate subject to radial compression. The strain hardening exponent, n, in expressions (41) is 12.

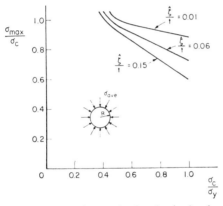

FIG. 6.   Maximum support stress for an elastic–plastic simply supported circular plate, when bifurcation takes place in the elastic range, for $n = 12$.

For an initial imperfection magnitude, in the shape of the critical mode, of 1% of the plate thickness, the stiffness reduction arising from plastic yielding leads to a maximum support stress, $\sigma_{\max}$, less than the elastic bifurcation stress, σ_c, when σ_c is greater than about 0.6 times the initial yield stress σ_y. The imperfection sensitivity increases as the bifurcation stress approaches the yield stress. A similar loss of load carrying capacity occurs for rectangular plates thick enough to have elastic bifurcation stresses approaching the yield stress. During the past 10–15 years a number of numerical analyses of the load carrying capacity of imperfect ideally plastic plates have been carried out (for example see references 34–6). These investigations have been directed toward determining the maximum support load of steel plating and some of these investigations have accounted for weld induced residual stresses.

A quite opposite phenomenon, which occurs for highly imperfection sensitive structures, when bifurcation of the perfect structure takes place near initial yielding, has been discussed by Hutchinson.[9] Hutchinson considered a long circular cylindrical shell subject to axial compression and made use of Koiter's[37] special solution for an axisymmetric imperfection. For a small imperfection, asymptotic estimates of the onset of initial yielding and of the maximum support load were made, assuming purely elastic material behaviour. Hutchinson[9] then considered the situation when the perfect shell bifurcates just at initial yield. If the maximum support load of the imperfect shell (assuming elastic material behaviour) is greater than about half the bifurcation load for the perfect shell, the maximum load is found to occur after plastic yielding. For larger imperfections, which give ratios of maximum load to critical load in the range 0·2–0·5, the load maximum is attained prior to initial yielding.[9] This is the range in which many thin cylindrical shells with typical imperfections buckle. Hutchinson[9] also notes that since the elastic analysis is valid for the perfect shell and in the range where the ratio of maximum support load to bifurcation load is between about 0·2 and 0·5, it is probably not far off for values of this ratio between 0·5 and 1·0. Furthermore, it is found that if the bifurcation load is about 95 % of that required for plastic yielding, the elastic analysis holds over the entire range likely to be of interest.[9]

When a structure which is highly imperfection sensitive in the elastic range undergoes bifurcation in the plastic range, both the decreasing stiffness for increasing deformation, associated with plastic deformation, and the geometric non-linearity are destabilising. The effects of this combination are illustrated in Hutchinson's[38] full numerical analysis of the axisymmetric post-bifurcation behaviour and imperfection sensitivity of a spherical shell subject to external pressure. Hutchinson[38] notes that the load reduction for a given imperfection magnitude is slightly less in a shell which undergoes bifurcation in plastic range than in a corresponding shell that bifurcates in the elastic range. Furthermore, shells that bifurcate in the plastic range will be thicker than shells that buckle purely elastically so that the magnitude of practical imperfection amplitudes as measured by the relevant ratio (imperfection amplitude to shell thickness) is apt to be smaller.

The examples cited so far have been for structures which have a unique buckling mode associated with the critical load. In a number of structures of practical interest there are multiple buckling modes associated with the critical load or there may be bifurcation loads only slightly higher than the critical load. For example, practical designs of stiffened panels often have a column mode and a local buckling mode available simultaneously or nearly simultaneously. Very little is known theoretically about mode interaction effects in the plastic range. Furthermore, relatively few numerical studies of mode interaction effects in the plastic range have followed the early pioneering investigation of Graves-Smith.[39]

A numerical study of the mode interaction behaviour of eccentrically stiffened elastic–plastic panels has been carried out by Tvergaard and Needleman.[40] The panels are taken to be infinitely wide and the stiffeners are represented as simple beams. Single bay panels and panels continuous over several bays in the axial direction are considered. Mode interaction leads to imperfection sensitivity in a single bay panel with positive mode deflections, where the skin is being further compressed by bending. For negative column mode deflections, where the skin is being stretched, considerable imperfection sensitivity, arising solely from the material non-linearity, is found. This effect of material non-linearity explains why the multi-bay panel is not less imperfection sensitive than the single bay panel, as it is in the elastic range.[41]

Figure 7, taken from reference 40, depicts a single bay panel designed so that elastic bifurcation occurs simultaneously in the wide column buckling

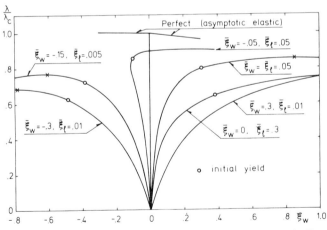

FIG. 7. Load versus wide column mode displacement for eccentrically stiffened panel supported at the two edges, with simultaneous elastic bifurcation in wide column and local buckling modes, for $n = 10$ and $-\sigma_c/\sigma_y = 0.909$.

mode and the local buckling mode. Here, ξ_w and $\bar{\xi}_w$ refer to the wide column mode amplitude and initial imperfection, respectively, while ξ_l and $\bar{\xi}_l$ denote corresponding local mode quantities. Linear elastic theory predicts that such a design is more imperfection sensitive than other designs, due to the strong interaction between the two critical modes. The asymptotic elastic post-bifurcation behaviour computed according to reference 3 is shown in this figure. For the panel depicted in Fig. 7 the critical stress is about 10 % below the yield stress. The panel is very sensitive to negative wide column mode initial imperfections even though a purely elastic panel is not. For positive column mode displacements, plasticity adds to the imperfection sensitivity. The behaviour exhibited in Fig. 7 is more complex than that depicted in Fig. 5, but the overall behaviour is similar.

A common observation in structures which buckle in the plastic range is that the final buckled configuration involves a highly localised pattern rather than a periodic one as assumed in many of the analyses discussed so far. For example, the final buckled configuration of a stiffened multi-bay panel may exhibit large deformations in one or a few bays rather than the periodic behaviour assumed in reference 40. Another example is provided by a steel plate strip where the final buckled configuration, as observed by Moxham[14] in over 100 tests, consists of one prominent buckle rather than the periodic pattern associated with the bifurcation mode. Moxham[42] used a model based on an assembly of single plates hinged together to argue that the continued growth of all buckles in a plate strip is not stable beyond the maximum load point. Recently, Tvergaard and Needleman[15] and Needleman and Tvergaard[10] have analysed buckling localisation for a wider class of structures. A simple one dimensional model was used to show that the basic mechanism of localisation is a bifurcation at the maximum load point, analogous to the necking of tensile bars. In more realistic structural models there is a delay between the maximum load point and the point of bifurcation leading to localisation, again analogous to the situation prevailing in tensile necking situations. The development of localisation was analysed numerically for several structural models including an elastic column on a softening foundation, an elastic–plastic column continuous over several equally spaced supports and a rectangular simply supported elastic–plastic plate. In all cases analysed it was found that it was the prelocalisation periodic deformation pattern which determined the maximum load point. Furthermore, small localised imperfections did not significantly effect the load carrying capacity of any of the structures analysed. These results indicate that the fact that localisation occurs does not necessarily invalidate an analysis directed toward determining the maximum support load which restricts attention to periodic modes and imperfections.

5. CONCLUDING REMARKS

Due largely to the work of Hill[5,6] a rather complete understanding of bifurcation in the plastic range has been achieved, for the classical smooth yield surface plasticity theories as well as for the more complex incremental plasticity theories, which permit vertex development on the yield surface. For smooth yield surface plasticity theory, in cases where there is a unique mode associated with the critical bifurcation load, rather general results characterising the initial post-bifurcation behaviour, due to Hutchinson,[8,9] are available. However, no rigorous relationship, analogous to that available in the elastic range,[1] has been established between the initial post-bifurcation behaviour of a perfect structure in the plastic range and the behaviour of a corresponding imperfect structure. A hypoelastic asymptotic analysis, which suppresses elastic unloading, has proven capable of revealing the main features of initial post-buckling behaviour and sensitivity to asymptotically small imperfections, both for the plate and shell structures analysed in references 12, 13 and for the columns analysed in the present paper. Comparisons with the results of full numerical solutions, including elastic unloading, indicate that the major limitation to the utility of such an analysis arises from its asymptotic nature.

Numerical investigations have played, and undoubtedly will continue to play, an important role in elucidating significant aspects of the post-buckling behaviour of plate and shell structures; for example, in the determination of the load carrying capacity of structures, which have a stable elastic post-bifurcation behaviour, in cases when bifurcation occurs sufficiently near yielding. Numerical solutions also play a prominent role in investigations of bifurcation from non-homogeneous stress states, where local plastic yield, say due to holes or cutouts, may precede bifurcation or shortly follow it (see, for example reference 43).

Mode interaction effects in the plastic range are not well explored. Little is known theoretically and relatively few numerical solutions have been obtained.

The highly localised final configuration often observed in structures that have failed due to plastic buckling has been explained as a bifurcation from a periodic to a localised mode subsequent to the attainment of a load maximum.

REFERENCES

1. KOITER, W. T., Over de stabiliteit van het elastisch evenwicht, *Thesis*, Delft, H. J. Paris, Amsterdam, 1945 (In Dutch). English translation: (a) *NASA TT-F10, 833*, 1967; (b) *AFFDL-TR-70-25*, 1970.
2. HUTCHINSON, J. W. and KOITER, W. T., *Appl. Mech. Rev.*, **23**, 1353–66, 1970.
3. TVERGAARD, V., *Proc. 14th IUTAM Congress*, W. T. Koiter, Ed., North-Holland, Amsterdam, 1976, pp. 233–47.
4. SHANLEY, F. R., *J. Aeronaut. Sci.*, **14**, 261–7, 1947.
5. HILL, R., *J. Mech. Phys. Solids*, **6**, 236–49, 1958.
6. HILL, R., *Problems of Continuum Mechanics*, Soc. Ind. Appl. Math., Philadelphia, Pennsylvania, 1961, pp. 155–64.
7. SEWELL, M. J., *Stability*, H. Leipholz, Ed. Univ. of Waterloo Press, Ontario, 1972, pp. 85–197.
8. HUTCHINSON, J. W., *J. Mech. Phys. Solids*, **21**, 163–90, 1973.
9. HUTCHINSON, J. W., *Advan. Appl. Mech.*, **14**, 67–144, 1974.
10. NEEDLEMAN, A. and TVERGAARD, V., in: *Mechanics of Solids, The Rodney Hill 60th Anniversary Volume*, H. G. Hopkins and M. J. Sewell, Eds., Pergamon Press, Oxford, 1981.
11. HUTCHINSON, J. W., *J. Mech. Phys. Solids*, **21**, 191–204, 1973.
12. NEEDLEMAN, A. and TVERGAARD, V., *Int. J. Solids Struct.*, **12**, 185–201, 1976.
13. TVERGAARD, V., *Int. J. Solids Struct.*, **13**, 957–970, 1977.
14. MOXHAM, K. E., Buckling tests on individual welded steel plates in compression. *Cambridge Univ. Eng. Dept. Report CUED/C-Struct/Tr.3*, 1971.
15. TVERGAARD, V. and NEEDLEMAN, A., *J. Appl. Mech.*, **47**, 613–19, 1980.
16. BUDIANSKY, B., *Advan. Appl. Mech.*, **14**, 1–65, 1974.
17. ONAT, E. T. and DRUCKER, D. C., *J. Aeronaut. Sci.*, **20**, 181–6, 1953.
18. HUTCHINSON, J. W. and BUDIANSKY, B., in: *Buckling of Structures*, B. Budiansky, Ed., Springer-Verlag, Berlin, 1976, pp. 98–105.
19. BATDORF, S. B., *J. Aeronaut. Sci.*, **16**, 405–408, 1949.
20. HILL, R., *J. Mech. Phys. Solids*, **14**, 95–102, 1966.
21. HUTCHINSON, J. W., *Proc. Roy. Soc., London A*, **318**, 247–72, 1970.
22. LIN, T. H., *Advan. Appl. Mech.*, **11**, 255–311, 1971.
23. MICHNO, M. J. and FINDLÈY, W. N., *Int. J. Non-Linear Mechanics*, **11**, 59–82, 1976.
24. HECKER, S. S., in: *Constitutive Equations in Viscoplasticity*, AMD-Vol. 20, ASME, New York, 1976, pp. 1–33.
25. SANDERS, J. L., in: *Proc. 2nd US Nat. Congr. Appl. Mech.*, Univ. of Michigan, Ann Arbor, Michigan, 1954, pp. 455–60.
26. SEWELL, M. J., *J. Mech. Phys. Solids*, **22**, 469–90, 1974.
27. CHRISTOFFERSEN, J. and HUTCHINSON, J. W., *J. Mech. Phys. Solids*, **27**, 465–87, 1979.
28. VAN DER HEIJDEN, A. M. A., *J. Mech. Phys. Solids*, **29**, 441–64, 1979.
29. TVERGAARD, V. and NEEDLEMAN, A., *Int. J. Mech. Sci.*, **17**, 419–24, 1975.
30. KOITER, W. T., Buckling and post-buckling behaviour of a cylindrical panel under axial compression, *National Luchtvaart Laboratorium 20, Report S476*, Amsterdam, 1956.

31. TVERGAARD, V., *Int. J. Solids Struct.*, **12**, 683–91, 1976; Errata, *ibid.*, **14**, 329, 1978.
32. CALLADINE, C. R., *Int. J. Mech. Sci.*, **15**, 593–604, 1973.
33. NEEDLEMAN, A., *Int. J. Mech. Sci.*, **17**, 1–13, 1975.
34. CRISFIELD, M. A., *Proc. Instn. Civ. Engrs.*, **59**, 595–624, 1975.
35. LITTLE, H. G., *Int. J. Mech. Sci.*, **19**, 725–44, 1977.
36. HARDING, J. E., HOBBS, R. E. and NEAL, B. G., *Proc. Instn. Civ. Engrs.*, **63**, 137–58, 1977.
37. KOITER, W. T., *Proc. Kon. Ned. Akad. Wetensch.*, **B66**, 265–79, 1963.
38. HUTCHINSON, J. W., *J. Appl. Mech.*, **39**, 155–62, 1972.
39. GRAVES-SMITH, T. R., *Thin-Walled Steel Structures*, Crosby Lockwood, London, 1969, pp. 35–60.
40. TVERGAARD, V. and NEEDLEMAN, A., *Int. J. Solids Structures*, **11**, 647–63, 1975.
41. KOITER, W. T. and PIGNATARO, M., *Buckling of Structures*, B. Budiansky, Ed., Springer Verlag, Berlin, 1976, pp. 133–48.
42. MOXHAM, K. E., Theoretical prediction of the strength of welded steel plates in compression, *Cambridge Univ. Eng. Dep. Report CUED/C-Struct/Tr.2*, 1971.
43. NEEDLEMAN, A., *J. AIAA*, **12**, 1594–6, 1974.

Chapter 7

NUMERICAL ANALYSIS OF STRUCTURES

M. A. CRISFIELD

*Structures Department,
Transport and Road Research Laboratory,
Crowthorne, Berkshire, UK*

SUMMARY

The chapter reviews the current state of the art in the numerical analysis of structures and gives a brief historical survey. Structural analysis covers so many areas that a complete review would require a book rather than a chapter. The author has therefore concentrated on those areas in which he has personal experience. As a result, some fields such as dynamics and fracture mechanics have been completely omitted and the paper concentrates on the non-linear analysis of thin-walled steel and concrete structures. Optimisation of structural form has meant that such structures are becoming thinner so that the collapse behaviour generally involves a complex interaction between material and geometric non-linearities. In such circumstances, numerical techniques are essential for a realistic structural analysis.

1. INTRODUCTION AND HISTORICAL BACKGROUND

Following a historical survey, the paper describes the main discretisation procedures of finite elements, finite differences and boundary integrals. Emphasis is placed on the finite element method which is today the dominant technique. The following sections on solution procedures relate to both linear and non-linear analysis. In recent years much research has been devoted to this topic and significant advances have been made. Such improvements are essential if non-linear numerical analysis is ever to be

235

economically viable for complex structures. The advances are needed as much to improve the robustness of the computer programs as to increase their speed. While reference is primarily made to the finite element method, most of the solution procedures are equally relevant to other discretisation techniques. In general, they are also applicable to both material and geometric non-linearities. However, material non-linearity introduces special problems that are described in a final section.

1.1. History

The foundation of structural analysis were laid at the end of the nineteenth century by Navier,[1] Maxwell,[2] Castigliano,[3] Lord Rayleigh,[4] Ritz[5] and others. At a similar time, the important mathematical tool of matrices was being developed.[6] Structural analysis and matrix methods were eventually wedded by the development of the digital computer in the early 1950s. Early applications involved frameworks and were pioneered by Livesley.[7] Before long, the important influence of minimum weight structures in the aircraft industry led to the development of element methods[8,9] which could be used to analyse continua as well as discontinua.

While progress has been made with finite differences, the finite element method[8,9,11] has become the dominant form of structural analysis. A history is given by Spooner.[12] The developments go back to the lattice analogy concepts developed by McHenry[13] and Hrenikoff[14] and early finite element work continued this framework analogy with the structure being envisaged as cut into elements. Before long the energy basis was recognised and more mathematical approaches were developed.[11,15] In particular, the finite element method was seen as an extension[15] of the Rayleigh–Ritz[4,5] procedure in which the integration of the energy was performed in a piece-wise manner using piece-wise continuous trial functions. The coefficients became 'nodal displacements'.

The importance of computers in the development of numerical analysis can be gauged from the observation of Jennings[16] that without machines it would take about one man-year to manually solve 40 linear simultaneous equations. The computer analysis of the Statfjord B off-shore platform involved 720 000 such equations.[17]

2. DISCRETISATION

Given a set of stresses σ, displacements $\mathbf{u}$ with prescribed values $\overset{*}{\mathbf{u}}$ on S, body forces $\mathbf{b}$ and external tractions $\mathbf{t}_e$ on S_σ, the following conditions need to be satisfied:

1. *Compatibility*:

Continuity and differentiability of $\mathbf{u}$ to the necessary degree (1)

Displacement boundary conditions, $\mathbf{u} = \overset{*}{\mathbf{u}}$ on S_u (2)

Strain–displacement law, $\varepsilon = \mathbf{Lu} + \varepsilon_n(\mathbf{u})$ (3)

where $\mathbf{L}$ is some linear operator and $\varepsilon_n(\mathbf{u})$ contains non-linear terms.

2. *Equilibrium*:

$$\mathbf{S\sigma} + \boldsymbol{\sigma}_n(\mathbf{u}, \boldsymbol{\sigma}) + \mathbf{b} = \mathbf{0} \tag{4}$$

$$\mathbf{t}_i = \mathbf{t}_e \quad \text{on } S_\sigma \tag{5}$$

where $\mathbf{t}_i$ are internal tractions, $\mathbf{S}$ is a linear operator and $\boldsymbol{\sigma}_n(\mathbf{u}, \boldsymbol{\sigma})$ contains non-linear terms.

3. *Stress–strain laws*:

$$\boldsymbol{\sigma} = \mathbf{E}(\mathbf{u}, \boldsymbol{\sigma})\varepsilon \tag{6}$$

For path dependent material, an equation of the form of eqn (6) may not exist. This phenomenon will be discussed later.

2.1. Finite Differences

Using the standard direct finite difference approach,[18] the differential operators $\mathbf{L}$ (eqn (3)) and $\mathbf{S}$ (eqn (4)) are replaced by difference operators, $\bar{\mathbf{L}}$ and $\bar{\mathbf{S}}$ involving discrete (nodal) values of $\mathbf{u}$ called $\mathbf{p}$ and discrete nodal values of $\boldsymbol{\sigma}$ called $\mathbf{s}$. Difference expressions are also set up for the non-linear terms. The equations may then be solved iteratively using a method such as dynamic relaxation (see sections on solution procedures). Alternatively, for linear systems at least, a direct solution may be used. Substitution of eqn (3) into eqn (6) and the resulting equation into eqns (4) and (5) would give an expression of the form

$$\bar{\mathbf{S}}\mathbf{E}\bar{\mathbf{L}}\mathbf{p} = \mathbf{Kp} = \mathbf{q} \tag{7}$$

where $\mathbf{q}$ are discretised nodal values of $\mathbf{b}$ and $\mathbf{t}_e$ in the same manner as $\mathbf{p}$ are nodal values of $\mathbf{u}$. Here $\mathbf{K}$ is the stiffness matrix which is generally non-symmetric; for while $\mathbf{S} = \mathbf{L}^T$, $\bar{\mathbf{S}} \neq \bar{\mathbf{L}}^T$. Interlacing meshes[19] have led to significant improvements in accuracy.

The major disadvantages of the finite difference method involve the difficulties of dealing with complex boundary conditions and irregular geometries which introduce irregular meshes. Both difficulties can be lessened by adopting an energy formulation.[20-22] If the linear total potential energy is

$$\Phi = a + \tfrac{1}{2}\int \varepsilon^T \mathbf{E}\varepsilon \, dv - \mathbf{q}^T \mathbf{p} \tag{8}$$

use of the difference operator

$$\varepsilon = \bar{\mathbf{L}}\mathbf{p} \tag{9}$$

and application of the principle of stationary potential energy,

$$\frac{\partial \Phi}{\partial \mathbf{p}} = \mathbf{0} \tag{10}$$

gives

$$\mathbf{q} = (\sum \bar{\mathbf{L}}^T \mathbf{E} \bar{\mathbf{L}})\mathbf{p} = \mathbf{K}\mathbf{p} \tag{11}$$

where $\mathbf{K}$ is the stiffness matrix. Not only does this approach lead to a symmetric stiffness matrix, but also the non-essential boundary conditions, eqn (5), do not have to be considered explicitly but are instead satisfied in a weighted average sense as a result of the stationary process (eqn (10)). Further advances on irregular meshes had been made by Pavlin and Perrone,[22] Liszka and Orkisz[23] and Mullord.[24] Finite difference methods have been applied to the non-linear analysis of both thin-walled steel plated structures[25,26] and shells.[27-9]

2.2. Finite Elements

The standard finite element formulation uses the principle of total potential energy and is a version of the Rayleigh-Ritz method with piece-wise continuous trial functions. The method can also be formulated using the principle of virtual work. The displacements $\mathbf{u}$ are related to nodal displacements $\mathbf{p}$ via piece-wise continuous trial functions (usually polynominals).

$$\mathbf{u} = \mathbf{N}\mathbf{p} \tag{12}$$

$\mathbf{N}$ are chosen to explicitly satisfy conditions (1) and (2). Equation (3) is also directly satisfied by differentiating eqn (12), i.e.

$$\varepsilon = \mathbf{L}\mathbf{u} + \varepsilon_n(\mathbf{u}) = \mathbf{L}\mathbf{N}\mathbf{p} + \varepsilon_n(\mathbf{N}\mathbf{p}) = \mathbf{B}_L\mathbf{p} + \varepsilon_n(\mathbf{p}) \tag{13}$$

Differentiation of eqn (13) gives

$$\mathrm{d}\varepsilon = (\mathbf{B}_L + \mathbf{B}(\mathbf{p}))\delta\mathbf{p} = \mathbf{B}\delta\mathbf{p} \tag{14}$$

where terms involving $\delta\mathbf{p}^2$ are neglected. Hence application of the principle of virtual work gives

$$\mathbf{q}^T\delta\mathbf{p} = \sum_e \int \boldsymbol{\sigma}^T \,\mathrm{d}\varepsilon\,\mathrm{d}v = \sum_e \int \boldsymbol{\sigma}^T \mathbf{B}\,\mathrm{d}v\,\delta\mathbf{p} \tag{15}$$

where the summation is carried out over all the elements. For (weighted average) equilibrium under any small virtual displacement $\delta\mathbf{p}$

$$\mathbf{q} = \mathbf{q}_i = \sum_e \int \mathbf{B}^T \boldsymbol{\sigma} \, dv \tag{16}$$

where $\mathbf{q}_i$ is the internal force vector. Equations (6), (13) and (16) can be solved, without being combined, using an iterative procedure such as conjugate gradients or dynamic relaxation. For a linear system, they can be combined to give a standard stiffness formulation. Substituting eqns (6) and (13) into eqn (16) (with $\mathbf{E}(\mathbf{u}, \boldsymbol{\sigma}) = \mathbf{E}$ and $\boldsymbol{\varepsilon}_n(\mathbf{p}) = \mathbf{0}$) gives

$$\mathbf{q} = \mathbf{Kp} = \left(\sum_e \int \mathbf{B}^T \mathbf{EB} \, dv \right) \mathbf{p} \tag{17}$$

(The subscript L on $\mathbf{B}_L$ will be dropped whenever purely linear systems are considered.) The integration of eqn (17) is often performed using numerical integration[30] with Gaussian quadrature being particularly popular for elements that can be mapped, using natural or curvilinear coordinates, to rectangles or rectangular prisms. For a two-dimensional element in the transformed ζ, η space.

$$\int\int \mathbf{B}^T \mathbf{EB} \, d\zeta \, d\eta = \sum_{j=1}^{N} w(j) \mathbf{B}^T(\zeta_j, \eta_j) \mathbf{EB}(\zeta_j, \eta_j) \tag{18}$$

where $w(j)$ are the weighting functions and ζ_j, η_j are the Gauss point positions. Using 2×2 Gaussian integration for the rectangle of Fig. 1, $N = 4$, $w(j) = 1$ for all j values and

$$\zeta_j = \frac{1}{\sqrt{3}}(-1, -1, 1, 1)$$

$$\eta_j = \frac{1}{\sqrt{3}}(-1, 1, -1, 1) \tag{19}$$

A more detailed discussion on the finite element method will be given later.

2.3. Boundary Integral Methods

Boundary integral techniques[31-34] use approximate functions that satisfy the governing equations in the domain (eqns (3) and (4)) but not on the

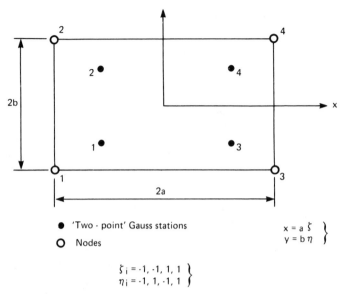

FIG. 1.   Linear rectangular element.

boundary. Consequently, there are generally less unknowns than will be generated by the finite difference or finite element methods. Other advantages involve the relative ease with which singularities and boundaries at infinity may be treated. For an application such as the analysis of a tunnel lining and adjacent soil, it may be useful to combine[32] the method with the finite element technique.

In comparison with the finite element method, the major disadvantage of boundary integral techniques would appear to relate to the added mathematical complexity. Also, the method leads to well populated rather than banded equations. Finally, the technique would appear less suitable for elasto-plastic problems since it is more difficult to find approximate functions that satisfy the governing equations in the domain. Nonetheless, advantages are still claimed for the method,[34] in particular a reduction in the required quantity of input data.

3.   FINITE ELEMENTS

Most linear finite element work has concentrated on the formation of the stiffness matrix since direct (Cholesky or Gauss) solution procedures are usually adopted. The stiffness matrix of eqn (17) was obtained via the principle of virtual work but the principle of stationary potential energy could equally have been used. The total potential energy method involves exact satisfaction of the compatibility relationships (1)–(3) and the stress–strain law, eqn (6), with a weighted average approximation of the equilibrium relationships, eqns (4)–(5). While the total potential energy method is the most straightforward, it is not necessarily the best. In particular, it might be better to spread the approximations rather than have them all relating to equilibrium. Indeed the satisfaction of the compatibility relationship (1) is sometimes very difficult. For the bending of thin plates, the 'necessary continuity' referred to in condition (1) involves the slopes $\partial w/\partial x$, $\partial w/\partial y$ being continuous between elements. This requirement follows from the inclusion of terms $\partial^2 w/\partial x^2$ in the potential energy (PE). For the piece-wise integration process to be valid, $\partial w/\partial x$ must be continuous. This has caused many problems and pure PE elements have generally been abandoned in this context. Similar arguments occur for the large-deflection Marguerre[35] type analysis of thin plates and beams. The strain ε_x is of the form

$$\varepsilon_x = \frac{\partial u}{\partial x} + \frac{1}{2}\left(\frac{\partial w}{\partial x}\right)^2 \tag{20}$$

With w a cubic in x (the minimum possible to ensure continuity of $\partial w/\partial x$), $\frac{1}{2}(\partial w/\partial x)^2$ is quartic in x so that, in general, if the constant strain mode, $\varepsilon_x =$ constant, is to be represented, u must be quintic in x!

This observation was made by Dawe.[36] While such high order elements are possible,[37] they are extremely unworkable in the context of irregular geometries, cut-outs, etc., and for efficient finite elements, modifications must be made to the pure PE approach.

The pure PE approach can be abandoned or modified in a number of ways. Firstly, it can be used with elements that do not exactly satisfy the continuity requirement (1). Such non-conforming elements may nonetheless converge and can be very successful.[38,39] Gallagher[41] and Thomas and Gallagher[42] have modified the PE approach by using Lagrangian multipliers to restore interelement compatibility. Other approaches involve adopting different variational principles.[11] In particular, a form of the Hu-Washizu variational principle[43] has strains and displacements as variables and allows both the strain-displacement laws, eqn (3), and the equilibrium conditions, eqns (4) and (5), to be satisfied in a weighted average sense as a result of the application of the variational principle. The Hellinger-Reissner principle[44] has stresses and displacements as variables and has the stress/differential-of-displacement relationships (combination of eqns (6) and (3)) as well as the equilibrium relationships satisfied in a weighted average sense.

Such variational principles can be applied directly but there is a general agreement (possibly because of an over-fixation with direct solution procedures) that generalised displacements should be the final variables so that a standard form of stiffness solution is eventually adopted. Since the finite element method is a piece-wise process, such hybrid procedures are possible with preliminary variations occurring at the element level. For example, with the hybrid stress method,[45-48] the complementary energy principle is used at the elemental level while the total potential energy approach applies at the structural level. An important application of this procedure is the Allman triangular plate bending element.[48]

Other, more heuristic approaches have been applied to modify the PE method. One very important practical method is reduced integration[49-53] in which the integration of the stiffness matrix (eqn (17)) is performed using a lower order numerical integration than that strictly required. The process is often applied in a selective manner. For example, the simple bi-linear plane stress rectangle has shape functions (eqn (12)) given by

$$u = \boldsymbol{\alpha}^T \mathbf{u} \qquad v = \boldsymbol{\alpha}^T \mathbf{v} \tag{21}$$

where

$$\alpha_i = \tfrac{1}{4}(1 + \zeta_i\zeta)(1 + \eta_i\eta) \tag{22}$$

and ζ_i and η_i takes the respective nodal values (Fig. 1) for $i = 1, 4$. Differentiation gives a strain matrix $\mathbf{B}_L$ (eqn (14)) for which two-point Gaussian integration is required if the stiffness matrix (eqn (17)) is to be correctly integrated. In particular, the shear strain is

$$\gamma_{xy} = \sum \tfrac{1}{4}\eta_i(1 + \zeta_i\zeta)u_i + \sum \tfrac{1}{4}\zeta_i(1 + \eta_i\eta)v_i \tag{23}$$

which represents the last line of eqn (14). Figure 2 shows the element being subjected to an important simple deformation state—pure bending. With linear displacement functions, a single element cannot exactly represent this state and the top and bottom edges remain straight. The deformations at the Gaussian integration stations are shown in the figure. Clearly, significant shear energy would be involved which would not exist in the real pure bending state. Consequently if few elements are used to model predominantly bending states, the resulting deformations will be far too small with significant energy having been involved in 'parasitic shear deformations'. Clearly, from Fig. 2, such parasitic shear deformations do not occur at the centroid. Consequently the solution can be improved by

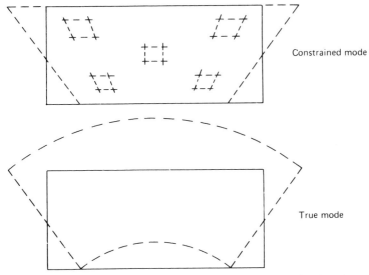

FIG. 2.   Parasitic shear strains in linear element subject to pure in-plane bending.

employing a selective form of reduced integration in which the ε_x and ε_y energy contributions to the stiffness matrix are integrated using two-point Gaussian integration while one-point integration is used for the shear energy. When directly applied, such selective schemes are computationally clumsy particularly for non-linear material problems in which the properties are stored at the Gauss points. Consequently a smoothed equivalent strain function approach can be adopted.[52,53] In the present case, this would involve replacing eqn (20) by its value at the one-point Gauss station (the centroid), i.e.

$$\gamma_{xy} = \sum \tfrac{1}{4}\eta_i u_i + \sum \tfrac{1}{4}\zeta_i v_i \qquad (24)$$

so that the last line of the **B** matrix (eqn (14)) or eqn (13) with $\varepsilon_n = 0$) is modified. Standard 2×2 Gaussian integration may now be used. (Similar approaches can be adopted for 'smoothing' other **B** functions to comply with 2×2 Gaussian integration.[52]) It is interesting that the stiffness matrix produced by this procedure can also be obtained using other more mathematical approaches such as the Hu-Washizu (with ε_{xy} = constant) or Hellinger-Reissner (with σ_{xy} = constant) variational principles.[53] In the context of plate bending, relationships between the Hellinger-Reissner variational principle and reduced integration are discussed in references 53 and 54.

Clearly, if such 'tricks' as reduced integration are to be adopted, methods must be established to check the validity of the resulting elements. Such a check has been developed by Irons[40,55] with the 'patch test' which is designed to test elements, however derived, using the computer. Irons argues that, in the limit, as the mesh is refined, a patch of elements will be required to exactly represent one of the constant strain states and should therefore be able to exactly represent these modes. Consequently prescribed displacements, that are consistent with a constant strain state, are applied (on the computer) to the boundaries of an arbitrary patch of elements. On solving the problem, the computer code should derive displacement and stresses everywhere in the patch that are consistent with the applied constant strain state. A mathematical analysis of the patch test is given by Strang and Fix.[56]

Reduced integration is frequently combined with 'Mindlin plate theory' (similar to Reissner's theory) which avoids the problems of continuous slopes between elements by omitting the Kirchhoff assumption and including shear deformation. Such techniques were originally advocated in relation to finite elements by Fraeijs de Veubeke[57] for beams in which the (linear) strain energy would be expressed as

$$\Phi = \frac{EI}{2} \int \left(\frac{\partial \theta}{\partial x}\right)^2 \mathrm{d}x + \frac{GA}{2} \int \left(\theta + \frac{\partial w}{\partial x}\right)^2 \mathrm{d}x \tag{25}$$

where θ is the rotation of the normal which is *not* (as with the normal Kirchhoff assumption) set to $-\partial w/\partial x$. If θ is set to $-\partial w/\partial x$, the energy would (as normal for thin plates) involve $\partial^2 w/\partial x^2$ and the shear energy would vanish. Using eqn (25) $\partial w/\partial x$ need not be continuous between elements since $\partial^2 w/\partial x^2$ does not appear. Hughes et al.[58] adopt simple linear expressions for w and θ. Reduced integration is required for the shear energy, if an excessively over-stiff solution is not to be obtained as the element gets thinner (or the shear modulus higher). In plate or shell terms, this method is equivalent to the approach of Ahmad et al.[59] which was significantly improved by the adoption of reduced integration.[44,50] At the present time the 'heterosis' quadrilateral element of Hughes et al.[58] is considered the most successful 'Mindlin plate'. Quadratic Lagrangian shape functions[11] are used for u and v while quadratic serendipity functions[11] (including a central node) are used for w. Selective integration is used for forming the stiffness matrix. Work is also in progress[61] on the development of a very simple element using linear displacements functions. At present the problem remains the avoidance of mechanisms.

The phenomenon of spurious mechanisms is primarily related to reduced integration but is also relevant to certain hybrid formulations. If the adopted order of numerical integration is too low, the stiffness matrix will be such that certain deformation patterns (other than rigid body modes) will involve zero strain energy and lead to a possible mechanism. The subject is not clear cut,[55] since although such mechanisms may be present for a single element, they may vanish when elements are put together. There is a further method for plate bending that starts with energy expressions including shear deformation. However in the discreet Kirchhoff[54,62,63] approach, the shear energy (second term in eqn (25)) is not actually included and the Kirchhoff|constraints (such as $\theta = -\partial w/\partial x$) are applied discreetly at various points (usually along the sides of the elements). In a recent paper on simple triangular bending elements (with three degrees of freedom at each corner node), Batoz et al.[54] considered the discreet Kirchhoff approach to be superior to both stress hybrid and reduced integration Mindlin formulations. The popular Irons' semi-loof shell element also involves discreet Kirchhoff constraints.[64]

3.1. Shells

A vast amount of literature has been produced on the application of finite elements to the analysis of thin shells. Until recently, the majority of this work has related to linear analysis. For example, a recent book (reference 65) is almost exclusively devoted to linear analysis and gives a good description of the state of the art. Gallagher has given reviews on both linear[41] and non-linear[66] shell analysis.

Some finite element formulations use coordinates embedded in the shell surface[36,37] and adopt deep shell theories such as those due to Koiter[68] and Budiansky and Sanders.[69] Alternatively, elements have been derived by modifying continuum elements to comply with shell assumptions without explicitly adopting a 'shell theory.[49,59,64] Such approaches are similar to the methods previously discussed which avoid the explicit adoption of the Kirchoff assumption in plate analysis. Extensions of the shell theory approach to non-linear analysis would involve a 'non-linear shell theory' such as that due to Koiter[70] or Sanders.[71] Such approaches have been applied using finite differences.[27-29] In contrast, most recent non-linear finite element formulations have extended the degenerate continuum concept[72-78] using either a total Langrangian[77] or up-dated Langrangian[74] approach.

Problems can be encountered when rotations are transformed from local to global coordinates. When the elements are planar, each element gives zero contribution to the diagonal stiffness for the in-plane rotation so that the stiffness matrix is singular. Fictitious rotational stiffnesses can be added to overcome this problem.[79] Alternatively, surface coordinates can be defined at the nodes using as normal, the average of the normals from the adjacent elements.[80] The in-plane rotation is not then defined and only two rotations are used in the surface coordinates. However for a general formulation which will handle complex jointed structures, the approach adopted in Irons' semi-loof element has great advantages.[64] In Irons' approach, the rotations are removed from the corners and are instead placed at 'Loof nodes' along the sides where only normal rotations are considered. The semi-loof element has been extended to non-linear situations by Javaherian et al.[75] and Dinis et al.[72] However it could be argued that non-linear analysis is better suited to a more simple element.

The earliest finite element solutions for linear shells adopted assemblages of flat or faceted elements.[79-81] Such formulations are simple and since rigid body motions are correctly represented,[67] the technique gives results that are only surpassed by the 'better' (and more complex) curved elements. There is currently a renewed interest in facet formulations[80,83] since they may be easily and economically introduced into a non-linear computer program using an up-dated coordinate system (co-rotational coordinates); good results have been obtained.[80,83] Such solution methods require good simple bending elements and this accounts for the recent renewal of interest in this topic.[54]

Special elements have been developed for shallow shells.[84-86] Although difficulties exist for a general formulation,[87] these shallow shell elements may be assembled to model a deep shell by introducing coordinate transformations.[86] It has been argued[87] that the resulting formulation will lead to a shallow shell solution rather than a true deep shell solution. However this should depend on the transformations and the nature of the adopted shallow shell theory. For instance, in a study of shallow and deep arches, Dawe[36] found that the use of Vlasov shell expressions,[88] which use surface coordinates, led to errors for deep arches. In contrast, the use of the Marguerre equations,[35] which depend on a flat reference surface, gave solutions that converged correctly to the deep arch solutions. In the limit, as the elements reduce in size, the Marguerre type shallow shell assemblages should coincide with a facet solution since the only difference between the formulations involves the slopes in the shallow shells relative to the facet which are the flat reference surfaces. As noted by Belytschko and Glaum,[89] when the elements get smaller, these slopes will tend to zero. Consequently, for fine meshes, there will be little difference between facet solutions and those that include such terms. Crisfield[53] has given an approach for the moderately large deflection analysis of cylindrical shells using such an assemblage of Marguerre type shallow shells.

4. SOLUTION PROCEDURES FOR LINEAR PROBLEMS

The solution procedures will sometimes be discussed in relation to the principle of minimum total potential energy although many of them can also be applied when other variational principles[11] are used. Indeed many methods may take as their starting point the virtual work equilibrium relationship of eqn (16). The advantage of introducing minimum total potential energy is that the structural solution procedures can then be related to the mathematical programming techniques developed in the field of unconstrained optimisation.[91-93] Before considering non-linear systems, a study will be made of solution procedures for linear problems. The total potential energy is

$$\Phi = a + \tfrac{1}{2}\mathbf{p}^T\mathbf{K}\mathbf{p} - \mathbf{q}^T\mathbf{p} \tag{26}$$

where the stiffness matrix $\mathbf{K}$ is given by eqn (17). Consequently at equilibrium,

$$\frac{\partial \Phi}{\partial \mathbf{p}} = \mathbf{K}\mathbf{p} - \mathbf{q} = \mathbf{g}(\mathbf{p}) = \mathbf{0} \tag{27}$$

Equation (27) can clearly be solved directly using Gaussian elimination or Cholesky decomposition.[94] Since K is generally well banded, banded solution procedures should be used. Alternatively, sky-line[95] or frontal methods[96] can be adopted.

For indirect solution procedures, trial values of $\mathbf{p}$ are adopted which are not at equilibrium so that

$$\mathbf{g}(\mathbf{p}) = \mathbf{q}_i(\mathbf{p}) - \mathbf{q} = \sum_e \int \mathbf{B}^T\boldsymbol{\sigma}(\mathbf{p})\,\mathrm{d}v - \mathbf{q} \tag{28}$$

where $\mathbf{q}_i$ comes from the virtual work eqns (15) and (16), i.e. given the displacement $\mathbf{p}$, the strain comes from eqn (14) and the stresses from eqn (6) and hence, from eqn (16), the internal forces $\mathbf{q}_i$, can be derived. This leads to the out of balance force vector $\mathbf{g}(\mathbf{p})$ or gradient of the total potential energy. Equation (28) is generally valid whether or not the system is linear. There are a number of mathematical programming methods for minimising Φ (or satisfying a set of simultaneous equations such as eqn (4)) while only using the gradient **g** and not the Hessian **K**. Such systems usually involve updating the trial displacements $\mathbf{p}_i$ according to

$$\mathbf{p}_{i+1} = \mathbf{p}_i + \eta_i \delta_i \qquad (29)$$

where δ_i is an iterative vector and the vector η_i is a 'step length' parameter.[91-93]

4.1. Conjugate Gradient Method[97-104]
From the conjugate gradient (cg) technique,

$$\delta_i = -\mathbf{g}_i + \beta_i \delta_{i-1} \qquad (30)$$

Using the Polak–Ribiere formula[99] which is generally considered to be better[100] than Hestenes-Steifel formula,[97]

$$\beta_i = \frac{\mathbf{g}_i^T(\mathbf{g}_i - \mathbf{g}_{i-1})}{\mathbf{g}_{i-1}^T \mathbf{g}_{i-1}} = \frac{\mathbf{g}_i^T \gamma_i}{\mathbf{g}_{i-1}^T \mathbf{g}_{i-1}} \qquad (31)$$

and

$$\eta_i \Rightarrow \min(\Phi(\mathbf{p}_{i+1})) \quad \text{or} \quad \eta_i \delta_i^T \mathbf{g}_{i+1} = 0 \qquad (32)$$

from which

$$\eta_i = \frac{-\delta_i^T \mathbf{g}_i}{\delta_i^T(\mathbf{g}_{i+1}(\eta = 1) - \mathbf{g}_i)} \qquad (33)$$

For linear problems, the conjugate gradient method is not strictly an iterative technique and will converge in n steps where n is the number of the equation. However, machine rounding errors mean that the technique effectively becomes an iterative procedure.

4.2. Scaled Conjugate Gradient Method
Although many improvements have been made[100-104] to the conjugate gradient method since its original formulation,[97] the method is still notoriously sensitive to rounding errors and has not been found very successful with structural applications. However the convergence characteristics can be dramatically improved by adopting a scaling procedure as advocated by Jennings and Malik,[102] Maierink and Vorst[103] and Kershaw.[104] The transformations,

$$\bar{\mathbf{p}} = \mathbf{L}^T \mathbf{p} \qquad \bar{\mathbf{q}} = \mathbf{L}^{-1} \mathbf{q} \qquad \bar{\mathbf{g}} = \mathbf{L}^{-1} \mathbf{g} \qquad (34)$$

allow the energy to be transformed to

$$\Phi = a + \tfrac{1}{2}\bar{\mathbf{p}}^T \bar{\mathbf{K}} \bar{\mathbf{p}} - \bar{\mathbf{q}}^T \bar{\mathbf{p}} \qquad (35)$$

where

$$\bar{K} = L^{-1}KL^{-T} \tag{36}$$

and L and L^T are the Cholesky factors of some approximate stiffness matrix K_a. The transformations of eqn (34) do not have to be explicitly applied and in the original space, eqn (30) may be replaced by

$$\delta_i = -K_a^{-1}g_i + \beta_i\delta_{i-1} \tag{37}$$

where eqn (31) is replaced by

$$\beta_i = \frac{-\overset{*}{\delta}{}_i^T\gamma_i}{\delta_{i-1}^T g_{i-1}} \tag{38}$$

$$\overset{*}{\delta}{}_i = -K_a^{-1}g_i \tag{39}$$

In references 102–104, such scalings were used in a 'partial elimination' method with K_a being the original stiffness matrix after some of the small off-diagonal terms had been neglected. (The Hestenes-Steifel formula was used rather than the Polak-Ribiere formula of eqn (31).) A rather simpler scaling was earlier adopted by Fox and Stanton[105] who set

$$K_a = D \tag{40}$$

where D is a diagonal matrix containing the diagonal terms of the stiffness matrix.

4.3. Dynamic Relaxation
Dynamic relaxation (DR)[106–109] is an iterative solution procedure of similar form to the conjugate gradient method. The name was originally derived as a dynamic analogy for solving static problems.[106] However it may alternatively be viewed as a second order Richardson process.[104–108] If scalings (eqn (34)) are first applied with

$$K_a = LL^T = D^{1/2}D^{1/2} = D \tag{41}$$

the procedure is of the same form as eqn (37) with

$$\eta = \frac{4}{(\sqrt{\gamma_m} + \sqrt{\gamma_0})^2}$$

$$\beta = \frac{(\sqrt{\gamma_0} - \sqrt{\gamma_m})^2}{4} \tag{42}$$

where γ_m and γ_0, respectively, are the maximum and minimum eigenvalues of the scaled stiffness matrix $\bar{K}$. Approximations are usually adopted with

Gerschgorin's theorem[94] being used for γ_m and possibly a coarse mesh lowest eigenvalue solution for γ_0. The method has been much used with finite differences,[107] but it should be stressed that the technique is independent of the adopted discretisation procedure and has also been used with finite elements.[109]

5. SOLUTION PROCEDURES FOR NON-LINEAR PROBLEMS

5.1. Incremental Methods
The simplest way to solve non-linear problems is to solve a set for incremental linear problems. Suppose that the loading term $\mathbf{q}$ (eqns (7), (11), (16) and (17)) is rewritten as $\lambda\mathbf{q}$ where the scaler λ represents some load level and $\mathbf{q}$ is a fixed load vector. Then the gradient of the total potential energy or out-of-balance force vector can be rewritten as

$$\mathbf{g}(\mathbf{p}, \lambda) = 0 \tag{43}$$

Given a solution at load level A, say $\mathbf{p}_A$, a Taylor's series expansion can be used to find

$$\mathbf{g}(\mathbf{p}, \lambda)|_B = \mathbf{g}(\mathbf{p}_A + \Delta\mathbf{p}, \lambda_A + \Delta\lambda) = \mathbf{g}(\mathbf{p}, \lambda)|_A + \frac{\partial \mathbf{g}}{\partial \mathbf{p}}\bigg|_A \Delta\mathbf{p} + \frac{\partial \mathbf{g}}{\partial \lambda}\bigg|_A \Delta\lambda$$

$$+ \frac{1}{2}\frac{\partial^2 \mathbf{g}}{\partial p_i \partial p_j}\bigg|_A \Delta p_i \Delta p_j + \frac{1}{2}\frac{\partial^2 \mathbf{g}}{\partial p_i \partial \lambda} \Delta p_i \Delta\lambda + \frac{1}{2}\frac{\partial^2 \mathbf{g}}{\partial \lambda^2}\Delta\lambda^2 + \cdots \tag{44}$$

If position A is at equilibrium,

$$\mathbf{g}(\mathbf{p}, \lambda)|_A = 0 \tag{45}$$

and if position B at $\lambda_B = \lambda_A + \Delta\lambda$ is also at equilibrium,

$$\mathbf{g}(\mathbf{p}_A + \Delta\mathbf{p}, \lambda_a + \Delta\lambda) = 0 \tag{46}$$

so that neglecting the higher order terms in eqn (44),

$$\frac{\partial \mathbf{g}}{\partial \mathbf{p}}\bigg|_A \Delta\mathbf{p} = \frac{-\partial \mathbf{g}}{\partial \lambda}\bigg|_A \Delta\lambda \tag{47}$$

But from eqn (28), with $\mathbf{q}$ replaced by $\lambda\mathbf{q}$,

$$\frac{\partial \mathbf{g}}{\partial \lambda} = -\mathbf{q} \tag{48}$$

$(\partial \mathbf{g}/\partial \mathbf{p})|_A$ is $\mathbf{K}_T$, the tangent stiffness matrix at A (which will be discussed), so that eqn (47) becomes

$$\Delta \mathbf{p} = \Delta \lambda \mathbf{K}_T^{-1} \mathbf{q} = \Delta \lambda \left. \frac{\partial \mathbf{g}}{\partial \mathbf{p}} \right|_A^{-1} \mathbf{q} \tag{49}$$

which is a pure incremental solution. Clearly the neglecting of the higher order terms in eqn (44) means that the solution is not exact. A set of solutions based on this technique would lead to the drift from equilibrium shown in Fig. 3.

A simple correction for the out of balance forces at level A can be made from eqns (44)–(48) via

$$\Delta \mathbf{p} = \mathbf{K}_T^{-1} (\Delta \lambda \mathbf{q} - \mathbf{g}(\mathbf{p}, \lambda)|_A) \tag{50}$$

Even adopting eqn (50), errors will accumulate and the incremental solution procedure should really be combined with some iterative correction method.

It should be noted that the tangent stiffness matrix, $\mathbf{K}_T$, need not be formed neither need eqn (49) be directly solved. Instead, eqn (46) can be solved iteratively for $\Delta \mathbf{p}$. If $\mathbf{p}_A$ and λ_a are held fixed and eqn (14) is used with

$$\Delta \boldsymbol{\varepsilon} = (\mathbf{B}_L + \mathbf{B}(\mathbf{p}_A)) \, \Delta \mathbf{p} \tag{51}$$

the incremental solution for $\Delta \mathbf{p}$ will involve the iterative solution of a set of linear equations which can be solved, as discussed before, using a technique such as DR or the cg method. However using this linearised version there is again a danger of a drift from equilibrium (Fig. 3). A major advantage of methods that involve the direct formation and factorisation of $\mathbf{K}_T$ is that a check can be kept on the stability of solution by monitoring the sign of the determinant of the stiffness matrix as the Cholesky factorisation is performed.

5.2. Tangent Stiffness Matrix
From eqn (28),

$$\frac{\partial \mathbf{g}}{\partial \mathbf{p}} = \sum_e \int \frac{\partial \mathbf{B}^T}{\partial \mathbf{p}} \boldsymbol{\sigma} \, \mathrm{d}V + \sum_e \int \mathbf{B}^T \frac{\partial \boldsymbol{\sigma}}{\partial \mathbf{p}} \, \mathrm{d}v \tag{52}$$

(For brevity, the $\sum_e$ term will now be discarded.) A more specific detailed exposition will show[11] that the first term in eqn (52) becomes $\mathbf{K}_G$, the so-called geometric stiffness matrix which is a function of the current stresses $\boldsymbol{\sigma}$. The second term can be expanded using eqn (14) to give

$$\int \mathbf{B}_L^T \mathbf{E} \mathbf{B}_L \, \mathrm{d}V + \int (\mathbf{B}(\mathbf{p})^T \mathbf{E} \mathbf{B}(\mathbf{p}) + \mathbf{B}_L^T \mathbf{E} \mathbf{B}(\mathbf{p}) + \mathbf{B}(\mathbf{p})^T \mathbf{E} \mathbf{B}_L) \, \mathrm{d}V \tag{53}$$

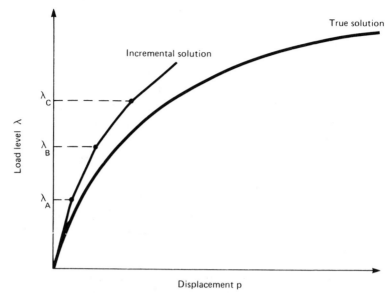

FIG. 3.   Drift from equilibrium with pure incremental solution.

The first term is the standard linear stiffness matrix, $\mathbf{K}_L$ (eqn (17)), while the second term is the initial displacement or initial slope matrix $\mathbf{K}_I$. Consequently,

$$\frac{\partial \mathbf{g}}{\partial \mathbf{p}} = \mathbf{K}_T = \mathbf{K}_L + \mathbf{K}_I + \mathbf{K}_G \qquad (54)$$

The initial slope matrix, $\mathbf{K}_I$, can alternatively be obtained by the up-dating of the coordinate transformations if an up-dated Lagrangian approach is adopted.[11] If material non-linearities are present $\mathbf{E}$ becomes $\mathbf{E}_T$. This point will be discussed in more detail later.

5.3. The Newton-Raphson Method

The most satisfactory way of solving non-linear problems is to combine incremental methods with iterative solution techniques. A number of such iterative methods will now be discussed.

Given any trial solution $\mathbf{p}$ for load level B, which is not in equilibrium, the out-of-balance force vector $\mathbf{g}(\mathbf{p}, \lambda_B)$ can be expanded as a Taylor series (with no change in λ) to give

$$\mathbf{g}(\mathbf{p} + \delta\mathbf{p}, \lambda_B) = \mathbf{g}(\mathbf{p}_B, \lambda_B) + \frac{\partial \mathbf{g}}{\partial \mathbf{p}}\bigg|_B \delta\mathbf{p} + \cdots \qquad (55)$$

so that an improved solution, which gives

$$g(\mathbf{p} + \delta\mathbf{p}, \lambda_B) = 0 \tag{56}$$

is obtained with

$$\delta\mathbf{p} = -\mathbf{K}_T^{-1}\mathbf{g}_B \tag{57}$$

and

$$\mathbf{K}_T = \frac{\partial\mathbf{g}}{\partial\mathbf{p}}\bigg|_B \tag{58}$$

being the 'tangent stiffness matrix' at $\mathbf{p}_B$, λ_B. Continued application, with no change in external load level, gives eqn (29) with $\eta_i = 1$, $\delta\mathbf{p}$ rewritten as $\boldsymbol{\delta}$ and

$$\boldsymbol{\delta}_i = -\mathbf{K}_i^{-1}(\mathbf{p}_i)\mathbf{g}_i \tag{59}$$

(The subscript T is dropped to make way for the subscript i, denoting the iteration number.) When combined with an initial tangential solution, the method gives the procedure shown in Fig. 4.

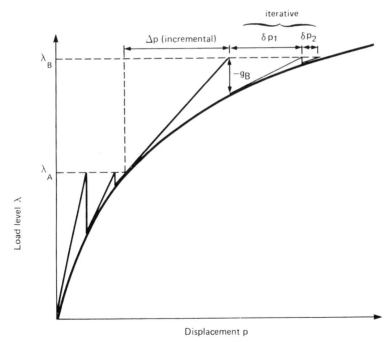

FIG. 4. Incremental procedure combined with Newton-Raphson iterations.

The Newton-Raphson method can have very fast convergence. However, it suffers from a number of drawbacks. In particular, the convergence can be very erratic near alternative unstable equilibrium points. Indeed K_i can become non-positive–definite since although K_i has been termed a tangent stiffness matrix, unless p is at equilibrium, it is not a true tangent stiffness matrix for the equilibrium path. Some improvements can be gained by introducing line searches so that $\eta_i \neq 1$. Using eqn (32), an optimum or near optimum 'step length' η_i may be found. A further drawback to the Newton-Raphson (N-R) method is the cost per iteration which can become excessive since K_i has to be both re-formed and re-factorised at each iteration.

5.4. The Modified Newton-Raphson Method

An obvious and important modification to the N-R method involves keeping K_i fixed at say K_a after a number of steps of N-R iterations. If K_i is held fixed at the true tangent stiffness matrix at the beginning of the increment, then with one variable, the technique is that shown in Fig. 5. At each iteration,

$$\delta_i = -K_a^{-1} g_i \qquad (60)$$

Clearly K_a (which is K_T at the beginning of the increment) need only be formed and factorised once per increment. As with the N-R method, convergence can often be improved or divergence prevented, by adding some form of line search. In reference 110, the crude damping procedure is adopted with η_i (eqn (29)) being given by

$$\eta_i = 0.8 \frac{\|g_i^T D^{-1} g_i\|}{\|g_{i+1}^T D^{-1} g_{i+1}\|} \qquad (61)$$

whenever

$$\|g_{i+1}^T D^{-1} g_{i+1}\| > 1.1 \|g_i^T D^{-1} g_i\| \qquad (62)$$

where D contains the diagonal terms of K_a. While this 'damping procedure' has been found satisfactory for problems involving the combined material and geometric non-linearity of steel structures, recent work by the author involving the cracking of concrete has indicated that more sophisticated 'line searches' may be required.[178]

With $K_a = $ true K_T, the modified Newton-Raphson method has the advantage that K_a will always be positive–definite provided the previous increment converged to a stable equilibrium state.

5.5. The Conjugate Gradient Method and Dynamic Relaxation

Both the cg method and DR can be used directly to solve non-linear

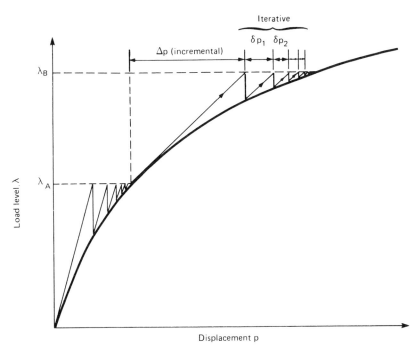

FIG. 5.   Incremental   procedure   combined   with   modified   Newton-Raphson
iterations.

problems. The major differences from the linear formulations involve the
formation of the gradient of out of balance force vectors **g**. Material and
geometric non-linearities are introduced via eqns (6) and (13), respectively.
With the cg method, the 'line search' of eqn (32) can no longer be simply
expressed using eqn (33). Alternative, more time-consuming methods are
necessary.[91–93] Also it is periodically necessary to 're-start' with a steepest
descent solution ($B_i = 0$ in eqn (30)). More sophisticated and efficient re-
start strategies are discussed in reference 100.

Simple scaling procedure would involve using eqn (4) with **D** containing
the diagonal stiffness coefficients of $\mathbf{K}_a$ at the beginning of the load
increment. In finite element terms, a more sophisticated scaling would
involve an approximate tangent stiffness matrix. Such an approach has
been developed by Irons and Elsawaf[111] and called a 'conjugate-Newton'
method. Crisfield's 'faster modified Newton-Raphson' iteration[110] or
secant-Newton methods[112] are also related to scaled cg techniques. With
the DR technique, the most significant changes from linear analysis involve

the possibility of changing the estimates of the maximum and minimum eigenvalues as the iterations proceed. The former can be achieved by periodic applications[25] of Gerschgorin's theorem[94] while an approximate form of the Rayleigh quotient[94] has been recommended[113,114] for updating the minimum eigenvalue. An alternative approach is given by Papadrakakis.[115] Dynamic relaxation has been extensively combined with finite differences for the non-linear analysis of thin-walled shells[28,29] and plated structures.[25,26]

5.6. The Variable Metric, Quasi-Newton Method

In the fields of mathematical programming and unconstrained optimisation, much work has been devoted to the development of quasi-Newton solution procedures.[91-93,116-120] Such methods resemble the N-R technique but do not require the explicit up-dating of the tangent stiffness matrix. Instead the stiffness matrix, or its inverse (or its Cholesky factors[91,118]) are continuously up-dated as the iterations proceed. That is, eqn (59) is adopted but $\mathbf{K}_i$ (or $\mathbf{K}_i^{-1}$) is up-dated iteratively from $\mathbf{K}_{i-1}$ (or $\mathbf{K}_{i-1}^{-1}$) to satisfy

$$\eta_{i-1}\delta_{i-1} = \mathbf{K}_i^{-1}\gamma_i \tag{63}$$

where

$$\gamma_i = \mathbf{g}_i(\mathbf{p}_{i-1} + \eta_{i-1}\delta_{i-1}) - \mathbf{g}_{i-1}(\mathbf{p}_{i-1}) = \mathbf{g}_i(\mathbf{p}_i) - \mathbf{g}_{i-1}(\mathbf{p}_{i-1}) \tag{64}$$

Equation (63) is known as the quasi-Newton or secant equation[91,118] which is exactly satisfied by a quadratic energy function.[91,93] A number of different up-dating formulae have been derived. The following so called BFGS formula,[119,120] is generally considered to be the best:

$$\mathbf{K}_i^{i-1} = \mathbf{K}_{i-1}^{-1} - \frac{\delta_{i-1}\gamma_i^T\mathbf{K}_{i-1}^{-1}}{\delta_{i-1}^T\gamma_i} - \frac{\mathbf{K}_{i-1}^{-1}\gamma_i\delta_{i-1}^T}{\delta_{i-1}^T\gamma_i} + \left(1 + \frac{\gamma_i^T\mathbf{K}_{i-1}^{-1}\gamma_i}{\eta_{i-1}\delta_{i-1}^T\gamma_i}\right)\frac{\eta_{i-1}\delta_{i-1}\delta_{i-1}^T}{\delta_{i-1}^T\gamma_i}$$

$$\tag{65}$$

The BFGS formula has a number of advantages. Firstly, provided certain conditions are met,[91,92,118] if $\mathbf{K}_{i-1}$ is positive–definite, the up-dated $\mathbf{K}_i$ will also be positive–definite. Secondly, the method is relatively insensitive to the accuracy of the line searches and can often be used with η_i (eqn (29)) $= 1$.

The quasi-Newton method has been used for structural analysis by a number of workers,[105,121-126] but in its original form it did not make much headway with the finite element method. Although the up-dates of eqn (65) can be applied directly to the Cholesky factors of $\mathbf{K}_i$ rather than to

K_i^{-1} (which is never explicitly formed in finite element work), the up-dates destroy the sparsity of these factors as they destroyed the sparsity of the original K matrix. Although Toint[127] and others have given methods for dealing with banded matrices, they appear somewhat cumbersome and the most important breakthrough for structural analysis, has been the work of Matthies and Strang[123] who altered the method to use vectors rather than explicitly up-dating K_i or K_i^{-1}. A similar one-step version of this vectorised approach was given at the same time by Crisfield.[110,112] Matthies and Strang have rewritten eqn (65) in the vector form

$$K_i^{-1} = (I + w_i v_i^T) K_{i-1}^{-1} (I + v_i w_i^T) \qquad (66)$$

(see reference 123 for the vectors v and w) and then substitute into eqn (59) so that

$$\delta_1 = -(I + w_1 v_1^T) K_0^{-1} (I + v_1 w_1^T) g_1 \qquad (67)$$

$$b_1 = g_1 + w_1^T g_1 v_1 \qquad (68)$$

$$c_1 = K_0^{-1} b_1 \qquad (69)$$

and

$$\delta_1 = -c_1 - v_1^T c_1 w_1 \qquad (70)$$

If the step length parameter η_i (eqn (29)) is unity (as it often will be), the procedure continues with

$$\delta_2 = -(I + w_2 v_2^T)(I + w_1 v_1^T) K_0^{-1} (I + v_1 w_1^T)(I + v_2 w_2^T) g_2 \qquad (71)$$

which can again be solved without ever changing K (or its Cholesky factors) from K_0. However, as the iterations proceed, more and more vectors are accumulated. This version of the BFGS method has been applied to structural analysis by Bathe and Cimento[124] and Abdel Rahman et al.[125] and excellent performances have been reported. Similar work on quasi-Newton methods has been done by Geradin et al.[126]

5.7. Secant-Newton Methods

In Crisfield's approach,[110,112] after each iteration, K is effectively re-set to K_0 so that eqn (71) would become

$$\delta_2 = -(I + w_2 v_2^T) K_0^{-1} (I + v_2 w_2^T) g_2 \qquad (72)$$

and the vectors do not continue to accumulate. In fact a slightly different formulation is adopted and the iterative change is

$$\delta_i = A_i \overset{*}{\delta}_i + B_i \eta_{i-1} \delta_{i-1} + C_i \overset{*}{\delta}_{i-1} \qquad (73)$$

where

$$\overset{*}{\delta}_i = -K_0^{-1} g_i \qquad \overset{*}{\delta}_{i-1} = -K_0^{-1} g_{i-1} \qquad (74)$$

are the present and past modified N-R changes (with fixed matrix $\mathbf{K}_0$). The scalar coefficients A_i, B_i and C_i change at iteration and involve inner products using $\mathbf{g}_i$, $\mathbf{g}_{i-1}$, $\boldsymbol{\delta}_{i-1}$, $\overset{*}{\boldsymbol{\delta}}_i$, $\overset{*}{\boldsymbol{\delta}}_{i-1}$. In one version, $C_i = 0$ and, in comparison with the standard modified N-R method, only two extra vectors $(\mathbf{g}_{i-1}, \boldsymbol{\delta}_{i-1})$ are required. For this procedure,[110,112]

$$A_i = \frac{-\boldsymbol{\delta}_{i-1}^T \mathbf{g}_{i-1}}{\boldsymbol{\delta}_{i-1}^T \boldsymbol{\gamma}_i}$$

$$B_i = \frac{-\boldsymbol{\delta}_{i-1}^T \mathbf{g}_i}{\boldsymbol{\delta}_{i-1}^T \boldsymbol{\gamma}_i} - A_i \frac{\overset{*}{\boldsymbol{\delta}}_i^T \boldsymbol{\gamma}_i}{\eta_{i-1} \boldsymbol{\delta}_{i-1}^T \boldsymbol{\gamma}_i} \tag{75}$$

In comparison to the standard modified N-R method, the cost at each iteration is only fractionally increased, yet the convergence characteristics are significantly improved (see Table 1). Equation (73) can also be thought of as a special form of scaled cg method which is related to an unusual cg method (with BFGS connections) due to Shanno.[101]

For one-dimensional problems both the 'secant-Newton' methods of eqn (73) and the BFGS quasi-Newton method coincide with the standard secant method. When combined with an incremental technique the procedure is that shown in Fig. 6.

TABLE 1

REQUIRED NUMBER OF ITERATIONS (EXCLUDING TANGENTIAL SOLUTION) FOR A FULLY CLAMPED CYLINDRICAL SHELL UNDER UNIFORM PRESSURE LOADING

Increment number	$q(\times 10^3 \, N \, mm^{-2})$	$\dfrac{w_c}{t}$	Solution method		
			Modified N–R	Eqn (75)	Eqn (73)
1	0·7	0·274	7	4	4
2	1·1	0·516	6	4	4
3	1·25	0·658	4	3	3
4	1·4	0·905	8	4	3
5	1·5	1·267	35	6	4
6	1·55	1·492	65	4	5
7	1·6	1·676	15	3	4
8	1·65	1·823	7	3	2
9	1·7	1·947	5	3	2
10	1·8	2·147	9	3	3
11	2·0	2·448	18	4	4

q: Young's modulus $= 3105 \, N \, mm^{-2}$, Poisson's ratio $= 0\cdot3$. Thickness, $t = 3175 \, mm$; radius $= 2540 \, mm$; length $=$ (arc) breadth $= 508 \, mm$. $w_c =$ central deflection.

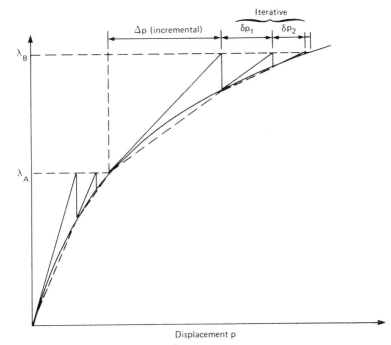

FIG. 6.   Incremental procedure combined with secant iterations.

5.8. Reduction and Perturbation Methods

Noor[128] has recently discussed a class of solution techniques involving a reduction in the number of active unknowns. The work of Nagy and König[129] could be considered in this category. They note that 'the true non-linear behaviour in the initial loading range can well be approximated by a linear combination of the eigenvectors of the linear buckling problem'. Clearly such approaches are related to Koiter's initial post-buckling methods.[130,131] The finite element formulations of Koiter's approach have also been given by Besseling,[132] Carnoy[133] and Haftka et al.[134] These procedures often involve perturbation techniques[135,136] in which the displacements $\mathbf{p}$ are expanded, using a Taylor's series, about some known solution ($\mathbf{p}_A$, λ_A), i.e.

$$\mathbf{p}_B = \mathbf{p}_A + \Delta\mathbf{p} \tag{76}$$

where

$$\Delta\mathbf{p} = \dot{\mathbf{p}}\,\Delta\lambda + \ddot{\mathbf{p}}\,\frac{\Delta\lambda^2}{2!} + \dddot{\mathbf{p}}\,\frac{\Delta\lambda^3}{3!} + \cdots \tag{77}$$

Substituting eqn (77) into eqn (44) and setting

$$\mathbf{g}(\mathbf{p}, \lambda)|_B = \mathbf{g}(\mathbf{p}, \lambda)|_A = 0 \qquad (78)$$

(i.e. exact equilibrium at the old and new positions), gives

$$0 = \left(\frac{\partial \mathbf{g}}{\partial \mathbf{p}}\bigg|_A \Delta \mathbf{p} + \frac{\partial \mathbf{g}}{\partial \lambda}\right) \Delta \lambda$$

$$+ \frac{1}{2}\left(\frac{\partial^2 \mathbf{g}}{\partial p_i \partial p_j}\bigg|_A \Delta \dot{p}_i \Delta \dot{p}_j + 2 \frac{\partial^2 \mathbf{g}}{\partial p_i \partial \lambda}\bigg|_A \Delta \dot{p}_i + \frac{\partial \mathbf{g}}{\partial \mathbf{p}}\bigg|_A \Delta \ddot{\mathbf{p}} + \frac{\partial^2 \mathbf{g}}{\partial \lambda^2}\bigg|_A\right) \Delta \lambda^2 + \cdots$$

$$(79)$$

For this equation to be satisfied for any load increase $\Delta \lambda$, the coefficients of $\Delta \lambda$, $\Delta \lambda^2$, etc., must each be zero so that

$$\Delta \dot{\mathbf{p}} = -\frac{\partial \mathbf{g}}{\partial \mathbf{p}}\bigg|_A^{-1} \frac{\partial \mathbf{g}}{\partial \lambda} = -\mathbf{K}_T^{-1} \mathbf{q}$$

$$\Delta \ddot{\mathbf{p}} = -\mathbf{K}_T^{-1}\left\{\frac{\partial^2 \mathbf{g}}{\partial p_i \partial p_j} \Delta \dot{p}_i \Delta \dot{p}_j + 2 \frac{\partial^2 \mathbf{g}}{\partial p_i \partial \lambda} \Delta p_i\right\} \qquad (80)$$

Consequently successive solutions for $\Delta \dot{\mathbf{p}}$ then $\Delta \ddot{\mathbf{p}}$ can be made which only require formation and factorisation of the tangent stiffness matrix at load level A. It is not necessary to use the load level as the perturbation parameter.[135] Instead some displacement variable or arc-length parameter (to be discussed) may be used.

Noor has recently presented an interesting combination of perturbation methods, Rayleigh-Ritz procedures and finite element methods.[128] The finite element and perturbation methods are used, as just described, to find the vectors $\Delta \dot{\mathbf{p}}$, $\Delta \ddot{\mathbf{p}}$. For the next few load increments, the Rayleigh-Ritz method is adopted with $\Delta \dot{\mathbf{p}}$, $\Delta \ddot{\mathbf{p}}$, etc., as trial functions, i.e. from equilibrium position A, any change in displacement is given by

$$\Delta \mathbf{u} = a_1 \Delta \dot{\mathbf{p}} + a_2 \Delta \ddot{\mathbf{p}} + a_3 \Delta \dddot{\mathbf{p}} + \cdots \qquad (81)$$

where a_1, a_2, etc., are the unknown coefficients which would number significantly less than the original equations. The reduced basis vectors $\Delta \dot{\mathbf{p}}$, $\Delta \ddot{\mathbf{p}}$, etc., are periodically recalculated when some error norm[128] exceeds a specified tolerance.

Other solution methods, that can be considered as using reduced bases, involve different levels of discretisation (possibly different meshes) at different stages in the iterative solution procedure.[112,137,138,193] An example, due to Crisfield,[53,112] involves the use of finite elements with hierarchical shape functions.[139] With the previously described secant-Newton iterative solution techniques, the matrix $\mathbf{K}_a$ can be any approximation to the true tangent stiffness matrix. By adopting hierarchical elements, the degrees of freedom can be partitioned as

$$\{\mathbf{p}\} = \begin{Bmatrix} \mathbf{p}_p \\ \mathbf{p}_s \end{Bmatrix} \tag{82}$$

where the primary displacement variables $\mathbf{p}_p$ involve a low order element (with $\mathbf{p}_s = 0$) and $\mathbf{p}_s$ are the additional relative nodal displacements required for the higher order element. Such a partitioning procedure leads to the following stiffness matrix

$$\mathbf{K} = \begin{bmatrix} \mathbf{K}_{pp} & \mathbf{K}_{ps} \\ \mathbf{K}_{ps}^T & \mathbf{K}_{ss} \end{bmatrix} \tag{83}$$

Since the primary variables represent a reasonable discretisation in their own right, the main energy will be associated with $\mathbf{K}_{pp}$. Consequently for an approximate stiffness matrix, $\mathbf{K}_{ps}$ can be neglected and $\mathbf{K}_{ss}$ can be represented by its leading diagonal terms only so that

$$\mathbf{K}_a = \begin{bmatrix} \mathbf{K}_{pp} & 0 \\ 0 & \mathbf{D}_{ss} \end{bmatrix} \tag{84}$$

This approximate stiffness matrix has the virtue of being well-banded, simply found and positive–definite. Applications show considerable promise.[112]

5.9. Automatic Load Incrementation

The choice of increment size is clearly important. If it is too large, convergence will not occur or will be very slow. Hence (neglecting for the moment path dependent material properties), the increment size should reflect the degree of non-linearity. A number of methods have been advocated for controlling the increment size.[140–143] In particular, Bergan et al.,[140,141] propose a current stiffness parameter as a measure of the degree of non-linearity. This parameter is a measure of the ratio of the present tangent stiffness to the tangent stiffness at the beginning of the loading. Bergan et al.[141] also note that the adoption of this procedure with the modified N-R method leads to a nearly constant number of iterations being required to achieve convergence at each increment. Consequently,

Crisfield[142] (using modified N-R or accelerated secant-Newton methods) adopted the single procedure

$$\Delta\lambda_j = \Delta\lambda_{j-1} \frac{I_d}{I_{j-1}} \tag{85}$$

to calculate the increment size. Here I_d is a desired number of iterations (approximately four) and I_{j-1} is the number of iterations required at a previous increment which was of size $\Delta\lambda_{j-1}$. Ramm[143] takes the square root of (I_d/I_{j-1}) which gives a smoother response. Clearly such a procedure would not work with iterative techniques such as cg or D-R. In such cases, Bergan's current stiffness parameter could be used.

5.10. Arc-Length Methods
An important technique for non-linear analysis, which has recently received considerable attention,[142,143,144] was developed by Riks[145,146] and Wempner[147] some 10 years ago. It involves incrementing the 'arc-length' of the solution path rather than the load or a specified displacement variable (Fig. 7). The method allows limit points to be passed automatically without having to change from load to displacement control or changing the displacement variable to be incremented. The method has also been found to be an efficient acceleration procedure.[143,144] Crisfield[112,144] and Ramm[143] have both recently presented versions of the technique which can be easily combined with the finite element method. In contrast to the original version, these techniques do not destroy the symmetry and banded nature of the governing equations. Ideas for further modifications to the method have been given by Riks.[146]

Figure 7 shows the method being applied to a one-dimensional problem using the modified N-R procedure. It will be noted that the load level λ changes as the iterations proceed and is therefore an additional variable. Since an additional variable has been added, an additional equation is required. Riks and Wempner proposed the constraint equation

$$\Delta\mathbf{p}_i^T \Delta\mathbf{p}_i + \Delta\lambda_i^2 \mathbf{q}^T\mathbf{q} = \Delta l^2 \tag{86}$$

where $\Delta\mathbf{p}$ is the incremental displacement, $\mathbf{q}$ is the total external load factor and Δl the prescribed 'arc length'. The constraint applies throughout the increment, for all iterations i. (Discussions on the choice of the prescribed arc length Δl are given in reference 144.) Crisfield[144] and Ramm[143] have found it better to use the simpler constraint

$$\Delta\mathbf{p}_i^T \Delta\mathbf{p}_i = \Delta l^2 \tag{87}$$

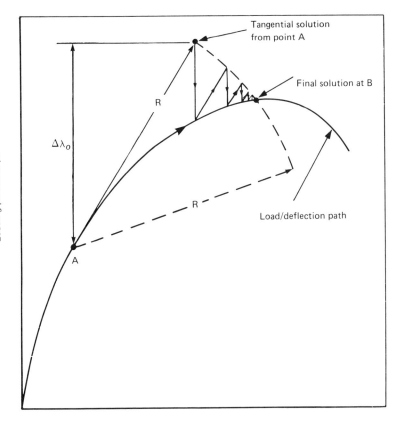

FIG. 7. Riks-Wempner arc length method with modified Newton-Raphson technique.

and Riks[145,146] has used a linearised version of eqn (86). The out-of-balance force vector $\mathbf{g}_i$ (eqn (28) with $\mathbf{q}$ replaced by $\lambda\mathbf{q}$) can be rewritten as

$$\mathbf{g}_i(\lambda_i + \delta\lambda_i) = \mathbf{g}_i(\lambda_i) - \delta\lambda_i\mathbf{q} \qquad (88)$$

Substituting into eqn (60), gives

$$\begin{aligned}\boldsymbol{\delta}_i(\lambda_i + \delta\lambda_i) &= -\mathbf{K}_a^{-1}\mathbf{g}_i(\lambda_i) + \delta\lambda_i\mathbf{K}_a^{-1}\mathbf{q} \\ &= \boldsymbol{\delta}_i(\lambda_i) + \delta\lambda_i\boldsymbol{\delta}_T\end{aligned} \qquad (89)$$

where $\boldsymbol{\delta}_i(\lambda_i)$ is the standard fixed load level (λ_i) modified N-R change, $\delta\lambda_i$ is the change in load level that will be required to satisfy the constraint

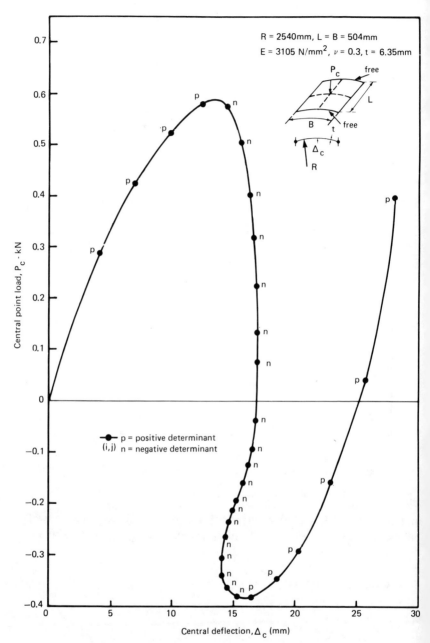

FIG. 8. Thin hinged cylindrical shell with a central point load (from reference 144).

eqn (87) and $\boldsymbol{\delta}_T$ is the tangent displacement which was found (and now stored) at the beginning of the load increment. To find the scalar $\delta\lambda_i$, eqn (89) (with $\eta_i = 1$) is substituted into eqn (87). This procedure leads to the simple scalar quadratic equation

$$a_1 \delta\lambda_i^2 + a_2 \delta\lambda_1 + a_3 = 0 \tag{90}$$

where a_1, a_2 and a_3 are given by

$$\begin{aligned}
a_1 &= \boldsymbol{\delta}_T^T \boldsymbol{\delta}_T \\
a_2 &= 2(\Delta\mathbf{p}_i + \boldsymbol{\delta}_i(\lambda_i))^T \boldsymbol{\delta}_T \\
a_3 &= (\Delta\mathbf{p}_i + \boldsymbol{\delta}_i(\lambda_i))^T (\Delta\mathbf{p}_i + \boldsymbol{\delta}_i(\lambda_i)) - \Delta l^2
\end{aligned} \tag{91}$$

The issue of choice of root and other details are discussed in reference 144. In comparison with the standard modified N-R method, only one extra vector $(\boldsymbol{\delta}_T)$ need be stored. The extra computation at each iteration is fairly negligible while not only are the convergence characteristics significantly improved but also difficult equilibrium paths can be traced as shown in Fig. 8. Such a path could not be traced using standard displacement control.

6. NON-LINEAR MATERIAL BEHAVIOUR

Non-linear material behaviour is easily handled by the finite element method[11,149] since a piece-wise integration scheme is involved. Using numerical integration, the tangent stiffness matrix is simply found by replacing $\mathbf{E}$ with $\mathbf{E}_T$ in eqns (15) or (53) where

$$\Delta\boldsymbol{\sigma} = \mathbf{E}_T \Delta\boldsymbol{\varepsilon} \tag{92}$$

$\mathbf{E}_T$ will generally be a function of the current stresses, and possibly strains, which must be stored at the (Gaussian) integration stations.

6.1. Plasticity
For the von Mises yield criterion, the explicit formulation of $\mathbf{E}$ was first given by Yamada et al.[150] and Zienkiewicz et al.[151] Under plane-stress, the yield criterion is

$$f = \frac{1}{\sigma_e}(\sigma_x^2 + \sigma_y^2 - \sigma_x\sigma_y + 3\tau_{xy}^2)^{1/2} = \frac{F^{1/2}}{\sigma_e} = 1 \tag{93}$$

where σ_e is the equivalent stress which, for a strain hardening material, is some function of the equivalent plastic strain

$$\varepsilon_{ps} = \sum \Delta\varepsilon_{ps} \tag{94}$$

$$\Delta\varepsilon_{ps} = \frac{2}{\sqrt{3}}(\Delta\varepsilon_{px}^2 + \Delta\varepsilon_{py}^2 + \Delta\varepsilon_{px}\Delta\varepsilon_{py} + \tfrac{1}{4}\Delta\gamma_{pxy}^2)^{1/2} \tag{95}$$

The incremental stress–strain laws are

$$\Delta\boldsymbol{\sigma} = \mathbf{E}(\Delta\boldsymbol{\varepsilon}_t - \Delta\boldsymbol{\varepsilon}_p) \tag{96}$$

while the Prandtl–Reuss flow rules are

$$\Delta\boldsymbol{\varepsilon}_p = \lambda^1 \frac{\partial f}{\partial\boldsymbol{\sigma}} = \lambda\frac{\partial F}{\partial\boldsymbol{\sigma}} = \lambda\bar{\boldsymbol{\sigma}} \tag{97}$$

where

$$\bar{\boldsymbol{\sigma}}^T = \{(2\sigma_x - \sigma_y),(2\sigma_y - \sigma_x),6\tau_{xy}\} \tag{98}$$

For the stresses to remain on the yield surface, differentiation of eqn (93) gives

$$\Delta f = \frac{1}{2\sigma_e^2}\bar{\boldsymbol{\sigma}}^T\Delta\boldsymbol{\sigma} - \frac{1}{\sigma_e}\frac{\partial\sigma_e}{\partial\varepsilon_{ps}}\Delta\varepsilon_{ps} = 0 \tag{99}$$

where $\partial\sigma_e/\partial\varepsilon_{ps}$ can be obtained (for a given ε_{ps}) from the uniaxial stress–strain curve. Substitution of eqn (97) into eqn (95) gives

$$\Delta\varepsilon_{ps} = 2\lambda\sigma_e \tag{100}$$

Pre-multiplying eqn (96) by $\bar{\boldsymbol{\sigma}}$ and, with the aid of eqns (97) and (100), substituting into eqn (99) gives

$$\lambda = \frac{1}{\left(\bar{\boldsymbol{\sigma}}^T\mathbf{E}\bar{\boldsymbol{\sigma}} + 4\sigma_e^2\dfrac{\partial\sigma_e}{\partial\varepsilon_{ps}}\right)}\bar{\boldsymbol{\sigma}}^T\mathbf{E}\,\Delta\boldsymbol{\varepsilon}_t \tag{101}$$

which must be positive or the stresses should be unloaded elastically from the yield surface. Using eqn (97) and substituting into eqn (96) gives:

$$\Delta\boldsymbol{\sigma} = \mathbf{E}_T\,\Delta\boldsymbol{\varepsilon}_t = \mathbf{E}\left(\mathbf{I} - \frac{1}{\left(\bar{\boldsymbol{\sigma}}^T\mathbf{E}\bar{\boldsymbol{\sigma}} + 4\sigma_e^2\dfrac{\partial\sigma_e}{\partial\varepsilon_{ps}}\right)}\bar{\boldsymbol{\sigma}}\bar{\boldsymbol{\sigma}}^T\mathbf{E}\right)\Delta\boldsymbol{\varepsilon}_t \tag{102}$$

Having used an incremental–iterative solution procedure to obtain the incremental strain $\Delta\varepsilon$ (the subscript t, for total, will be dropped where no confusion will arise), the incremental stresses can be obtained using

$$\Delta\sigma = \mathbf{E}_T(\sigma_0)\,\Delta\varepsilon \qquad (103)$$

where σ_0 are the stresses at the beginning of the increment. Alternatively, with less computation (unless $\mathbf{E}_T$ is stored), eqns (101), (97) and (96) may be used directly.

$$\sigma_n = \sigma_0 + \Delta\sigma \qquad (104)$$

will generally lie outside the yield surface (at point B, Fig. 9(a)). It is important that the stresses are not allowed to drift outside the yield surface since the error will generally accumulate as the increments are combined, particularly for elastic–perfectly plastic materials. An easy way to bring the stresses to the yield surface, is to apply further artificial plastic strain with no further total strain increases. (The yield function is known with more accuracy and is more important than the flow rules.[11]) Hence, given a required shift Δf in the yield function (to return it from f_0 to unity), a similar operation to that for eqn (101) gives

$$\lambda_a = \frac{-2f_0\sigma_e^2\,\Delta f}{\bar{\sigma}^T\mathbf{E}\bar{\sigma} + 4\sigma_e^2 f_0^2 \dfrac{\partial\sigma_e}{\partial\varepsilon_{ps}}} \qquad (105)$$

With $\Delta\varepsilon_t$ now zero, eqns (95) and (96) can be used to find the changes in stress to bring the yield function back to unity (Fig. 9(a)).

6.2. Sub-Increments

Particularly if large load increments are adopted, it is more advisable to integrate (probably numerically) the flow rules using a sub-incremental technique[11,152] (Fig. 9(b)). At each Gauss point, the incremental strain $\Delta\varepsilon$ is sub-divided into $n =$ sub-increments $(1/n)\,\Delta\varepsilon$. Knowing the 'devioteric stresses' at the beginning of the increment, $\bar{\sigma}_s$ (eqn (98)), eqns (101), (97) and (90) can be used to find the tangential $\Delta\sigma_s$. If necessary a shift can then be applied using eqn (105). New starting devioteric stresses $\bar{\sigma}_{s+1}$ are then found (Fig. 9(b)). Clearly the sub-incremental strategy involves less departures from the yield surface. The issue of integrating the flow rules is discussed by Key et al.[113] and Bushnell.[153]

For problems involving combined material and geometric non-linearity, Bushnell adopts a somewhat unusual solution procedure.[153] The material

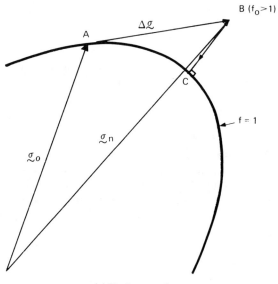

(a) Single correction

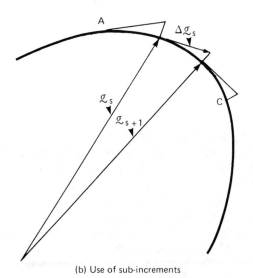

(b) Use of sub-increments

FIG. 9.   Keeping the stresses on the yield surface.

properties are held fixed while iterations are performed (in Bushnell's case using the N-R procedure) on the geometric non-linearities to obtain equilibrium. At this stage the sub-incremental procedure is used to integrate the flow rules and obtain an adjusted set of stresses. Using the new (again fixed) material properties at the end of the increment, the out of balance forces are again dissipated by means of N-R iterations. A simple one-step version of this procedure (using the strategy of Fig. 9(a) rather than that of Fig. 9(b)) is adopted by Little[121] using the quasi-Newton iterative technique. These procedures differ from the usual method in which the iterations are performed to simultaneously account for both material and geometric non-linearities. The advantage of the separation is that sophisticated solution procedures (quasi-Newton or Newton) can be used to their maximum efficiency with a smooth governing function (PE) without material changes. Consequently large increments can be applied. However, it should be noted that the procedures change the real path of the solution. This could matter for path dependent material behaviour (such as plasticity). Indeed, as noted by Argyris et al.,[154] the standard sub-incremental procedure also leads to violations. For, however closely the yield function may be followed and the flow rules integrated, the system is only in equilibrium at the beginning and end of the increment. The extent to which such path violations matter is open to debate. For imperfect steel plates subject to uniaxial compression, Little[121] has shown that accurate answers can be obtained with very large increments. An adaptive incremental procedure for elasto-plastics problems is proposed by Tracey and Freese.[155]

6.3. Plates and Shells

Early approaches to elasto-plastic plates involve tracing the elasto-plastic interface.[156] However this approach has generally been superseded by a layered or volume integral approach in which the stiffnesses are found by numerically integrating through the depth.[157-9] At each layer (or integration level) the previously discussed plane-stress relationship is assumed to apply.

Plane sections are assumed to remain plane so that

$$\Delta \varepsilon_z = \Delta \varepsilon_0 + z \, \Delta \chi \qquad (106)$$

where ε_0 is the strain at the reference plane (not necessarily the centre line). With the incremental stress resultant defined by

$$\Delta N = \int \Delta \sigma \, dz \qquad \Delta M = \int z \, \Delta \sigma \, dz \qquad (107)$$

substitution of eqns (92) and (106) into eqn (107) gives

$$\Delta N = C \Delta \varepsilon + G \Delta \chi$$
$$\Delta M = G^T \Delta \varepsilon + D \Delta \chi \qquad (108)$$

where

$$C = \int E_T \, dz \qquad G = \int z E_T \, dz \qquad D = \int z^2 E_T \, dz \qquad (109)$$

A number of different procedures have been used for integrating through the depth. Marcal and Pilgrim[157] used the trapezoidal rule with 11 integration stations while others have used the simple mid-point rule (separate layers). Gaussian integration was originally avoided since it does not have any integration stations on the surfaces. Crisfield[158] argued that it was an integrated stiffness that mattered and that failure to have points on the surface would not be very significant. Consequently he adopted Gaussian integration (generally with five integration stations) when analysing imperfect steel plates involving both material and geometric non-linearities. In a comparison of different integration techniques, Cormeau[159] favoured Gaussian integration.

The 'layered' approach to plasticity leads to considerable demands on both computer storage and time. These demands can be relieved by adopting a yield criterion that is a direct function of the stress resultants (N and M).[158,160-65] Figures 10 and 11 shows results obtained in this manner[142,155] when computing the collapse behaviour of a steel box-girder bridge. Both material and geometric non-linearities were involved.

6.4. Concrete

Two of the earliest non-linear analysis of reinforced concrete structures were due to Scordelis[166] and Nilson[167] who followed cracking by separating the nodes and using tie links to simulate bond. More recent work involving discrete cracks is given by Blauwendraad et al.[168] In general, analysis procedures have favoured the smeared property approach[169] which is effectively similar to that previously described for plasticity in steel. Smeared equivalent stress and strain relationships (either tangential or secant) are obtained at the numerical integration points. These relationships allow for cracking, crushing, yielding, etc. Many different approaches have been adopted (see the reviews by Bergan and Holand,[170] Bazant[171] and Argyris et al.[172]).

For slabs[173-8] and shells,[175,176] both non-linear elastic[173] and elasto-plastic[174,175] formulations have been used. Slabs are often under reinforced and in such cases undue sophistication in the compression zone would seem

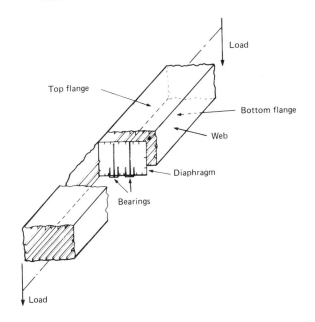

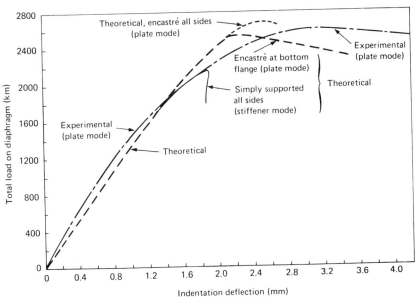

FIG. 10. Relationship between load and indentation deflexion at bearing for stiffened box-girder.

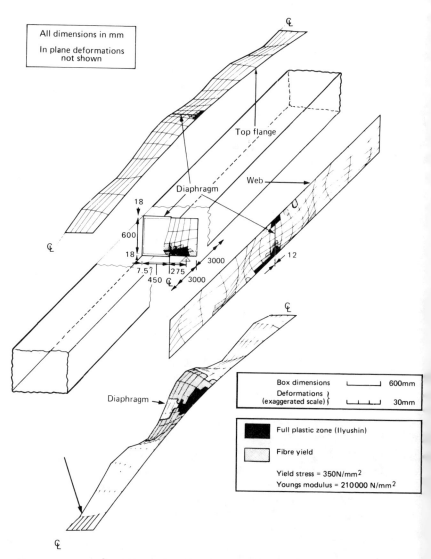

FIG. 11. Predicted deformations and plastic zones at collapse of steel box-girder (in-plane deformations not shown).

unwarranted since cracking is the most important phenomenon. The following basic procedure, which is similar to many other formulations,[174,175] has been adopted by the author to find $\mathbf{E}_T$ (see Fig. 12). At each integration point, the in-plane stresses are resolved into σ_1, σ_2, τ_{12} (Fig. 12) via

$$\begin{Bmatrix} \sigma_1 \\ \sigma_2 \\ \tau_{12} \end{Bmatrix} = \begin{bmatrix} c^2 & s^2 & -2sc \\ s^2 & c^2 & 2sc \\ sc & -sc & (c^2 - s^2) \end{bmatrix} \begin{Bmatrix} \sigma_x \\ \sigma_y \\ \tau_{xy} \end{Bmatrix} = \mathbf{T}\boldsymbol{\sigma}_{xy} \qquad (110)$$

with

$$s = \sin \theta \qquad c = \cos \theta$$

where θ is either the direction of the principal stresses (with $\tau_{12} = 0$), for first cracking, or else the previously recorded crack direction. In the 12 space, the 'yield criterion' of Fig. 13 is adopted. In the compresssion–compression zone, the von Mises yield criterion is used. This gives a fairly close fit to

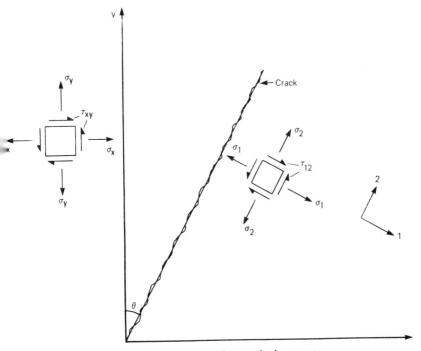

FIG. 12. Stress systems for cracked concrete.

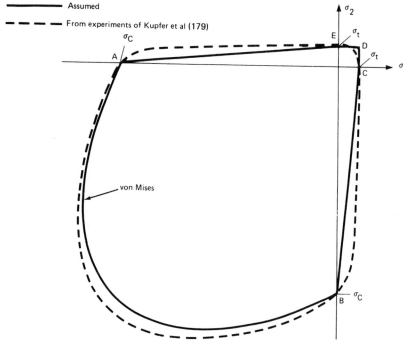

FIG. 13.   Concrete failure criterion.

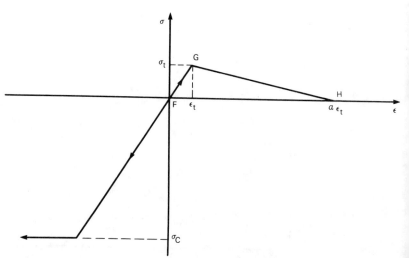

FIG. 14.   Assumed uniaxial stress–strain relationship for concrete.

the experimental results of Kupfer *et al.*[179] (Fig. 13). Although strain softening can be included in the compression zone,[170-172] this sophistication is not considered necessary for under reinforced thin slabs. If the stresses lie on the lines BC, CD, DE or EA (Fig. 13), cracking is assumed to occur and the direction θ is fixed (and stored). The strain softening approach of Fig. 14 is adopted to account for 'tension stiffening' in the tensile zones.[173,175] This tension stiffening allows for the growth of tensile

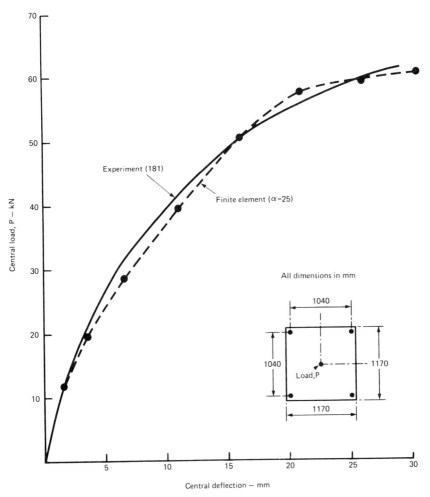

FIG. 15. Load–deflection relationship for Braunschweig corner supported slab.

stresses in the concrete of a reinforced concrete structure between cracks. After cracking, Poisson's ratio is assumed to be zero. To complete the definition of the stiffnesses in the two directions (number 1 and number 2), the shear modulus is set to a reduced factor ($\simeq 0.6$) of the full shear modulus[176] and $\mathbf{E}_T$ can be found in the original x–y directions via

$$\mathbf{E}_{T(x,y)} = \mathbf{T}_\varepsilon^{-1} \mathbf{E}_{T(1,2)} \mathbf{T}_\varepsilon^{-T} \tag{111}$$

where $\mathbf{T}_\varepsilon$ is the strain transformation matrix which, with an 'engineering' definition for the shear strains, is given by

$$\begin{Bmatrix} \varepsilon_1 \\ \varepsilon_2 \\ \gamma_{12} \end{Bmatrix} = \begin{bmatrix} c^2 & s^2 & -sc \\ s^2 & c^2 & sc \\ 2sc & -2sc & (c^2 - s^2) \end{bmatrix} \begin{Bmatrix} \varepsilon_x \\ \varepsilon_y \\ \gamma_{xy} \end{Bmatrix} = \mathbf{T}_\varepsilon \varepsilon_{xy} \tag{112}$$

More sophisticated tension stiffening models have been used.[173,177] In particular, the tension stiffening may decrease as the distance from the steel increases.

The sharper the falling line GH in Fig. 14 (i.e. the less the tension stiffening), the more difficult is the numerical analysis.[178] The point may be reached where the tangent stiffness matrix for the structure is not positive–definite and non-unique solutions may arise.[178,180] The phenomena may not be detected by some analysis which use the 'initial stress' approach[11] in which the stiffness matrix is not up-dated so that no check is made on the stability of the solution. Recently, more sophisticated numerical procedures such as the BFGS, arc-length and secant-Newton methods have been used for concrete slab analyses.[174,178] Figure 15 shows a comparison between theory[178] and experiment.[181]

7. CONCLUSIONS

The finite element method is today the most popular technique for the numerical analysis of structures. The main reason for this dominant position is the ability of the method to handle complex geometries without resorting to special procedures. Hughes[182] has observed that the year 1970 saw a reduction in the published literature on finite elements and concluded that this decline coincided with the decline in aerospace activity. He also noted that since 1970 research work has flourished, indicating the growth of interest in finite elements in other fields of engineering, such as off-shore structures.

The development of faster and cheaper computers has led to greater emphasis being placed on non-linear collapse analyses rather than linear 'stress at a point' calculations. This latter activity has coincided with the development of new codes of practice embodying limit state concepts which require direct calculations for the collapse limit state. Numerical techniques have been extensively used in the development of such codes. This activity is likely to continue and indeed increase, since recent developments in non-linear analysis are dramatically improving the speed and, of more importance, the robustness of non-linear computer programs. With computers becoming cheaper and faster, direct application in industry will also increase. The most important danger in these developments involves the educational gap between the researchers who develop the techniques and computer programs and the engineers who will use them.

REFERENCES

1. NAVIER, H. M. C., 'Resumé des leçons . . . de la résistance des corps solides', Paris, 1826. (Distributed as lecture notes in 1819.)
2. MAXWELL, J. C., *Phil. Mag.*, **27**, 294–9, 1864.
3. CASTIGLIANO, A., 'Théorie de l'équilibre des systèmes élastiques', Turin, 1879.
4. STRUTT, J. W. (LORD RAYLEIGH), *Trans. Roy. Soc. (London)*, *A*, **161**, 77–118, 1870.
5. RITZ, W., *J. Reine angew Math.*, **135** 1–61, 1909.
6. CAYLEY, A., *London Phil. Trans.*, **148**, 17–37, 1857.
7. LIVESLEY, R. K., *Engineering*, **176**, 230, 1953.
8. ARGYRIS, J. H. and KELSEY, S., 'Energy theorems and structural analysis', *Aircraft Engineering*, October 1954–May 1955. (Reprinted by Butterworths, London, 1980.)
9. TURNER, M. J., CLOUGH, R. W., MARTIN, H. C. and TOPP, L. J., *J. Aero. Sci.*, **23**, 805–23, 1956.
10. SOUTHWELL, R. V., *Relaxation Methods in Theoretical Physics*, Clarendon Press, Oxford, 1946.
11. ZIENKIEWICZ, O. C., *The Finite Element Method*. 3rd Edn, McGraw-Hill, Maidenhead, UK, 1977.
12. SPOONER, J. B., *World Congress on Finite Element Methods in Structural Mechanics*, Vol. 1, J. Robinson, Ed., Bournemouth, 1975, pp. A1–A22.
13. McHENRY, D., *J. Inst. Civ. Eng.*, **21**, 59–82, 1943.
14. HRENIKOFF, A., *J. Appl. Mech.*, **A8**, 169–75, 1941.
15. JONES, R. E., *J. AIAA*, **2**, 821–6, 1964.
16. JENNINGS, A., 'A giant's stride in engineering', Inaugural lecture, Queen's University of Belfast, November 1978.
17. HARWISS, T., *Finite Element News*, 6–9, September 1979.
18. FORSYTHE, G. E. and WASOW, W. R., *Finite Difference Methods for Partial Differential Equations*, Wiley, New York, 1960.

278   M. A. CRISFIELD

19. GILLES, D. C., *Proc. Roy. Soc. Lond.*, 407–33, 1948.
20. NOOR, A. K. and SCHNOBRICH, W. C., in: *Variational Methods in Engineering*, Vol. 2, C. A. Brebbia and H. Tottenham, Eds., Southampton University Press, Southampton, 1973, pp. 12/1–12/50.
21. BUSHNELL, D., ALMROTH, B. O. and BROGAN, F., *J. Comp. and Struct.*, 3, 361–87, 1971.
22. PAVLIN, V. and PERRONE, N., *Int. J. Num. Meth. in Engng.*, 14, 647–64, 1979.
23. LISZKA, T. and ORKISZ, J., *J. Comp. and Struct.*, 11, 83–95, 1980.
24. MULLORD, P., *Appl. Math. Modelling*, 3, 433–40, 1979.
25. FRIEZE, P. A., HOBBS, R. E. and DOWLING, P. J., *J. Comp. and Struct.*, 8, 301–10, 1978.
26. HARDING, J. E. and HOBBS, R. E., *The Structural Engineer*, 57B, 49–54, 1979.
27. BUSHNELL, D., *Int. J. Solids and Struct.*, 6, 157–81, 1970.
28. HARDING, J. E., *Proc. Instn. Civ. Engrs.*, 65(2), 875–92, 1978.
29. FRIEZE, P. A. and SACHINIS, A., in: *Numerical Methods for Non-Linear Problems*, C. Taylor *et al.*, Eds., Pineridge, Swansea, 1980, pp. 367–79.
30. IRONS, B. M., *Conf. on Use of Digital Computers in Struct. Eng.*, University of Newcastle, 1966.
31. BREBBIA, C. A., *The Boundary Element Method for Engineers*, Pentech, Plymouth, 1978.
32. ZIENKIEWICZ, O. C., KELLY, D. W. and BETTES, P., *Int. J. Num. Meth. in Engng.*, 11, 355–76, 1977.
33. TELLES, J. C. F. and BREBBIA, C. A., in: *Nonlinear Finite Element Analysis in Structural Mechanics*, W. Wunderlich *et al.*, Eds., Springer Verlag, Berlin, 1981, pp. 403–34.
34. CATHIE, D. N., *Appl. Math. Modelling*, 5, 39–48, 1981.
35. MARGUERRE, K., in: *Proc. 5th Int. Congress Appl. Mech.*, Wiley, New York, 1938, pp. 93–101.
36. DAWE, D. J., *Comp. and Struct.*, 4, 559–82, 1974.
37. DAWE, D. J., in: *Finite Elements for Thin Shells and Curved Members*, D. G. Ashwell and R. H. Gallagher, Eds., Wiley, London, 1976, pp. 131–53.
38. TAYLOR, R. L., BERESFORD, P. J. and WILSON, E. L., *Int. J. Num. Meth. in Engng.*, 10, 1211–20, 1976.
39. ZIENKIEWICZ, O. C. and CHEUNG, Y. K., *Proc. Instn. Civ. Engrs.*, 28, 471–88, 1964.
40. IRONS, B. M. and RAZZAQUE, A., in: *Variational Methods in Engng.*, Vol. 1, 1–73, C. Brebbia and H. Tottenham, Eds., Southampton University Press, Southampton, 1973, pp. 4159–72.
41. GALLAGHER, R. H., in: *World Congress on Finite Element Methods in Structural Mechanics*, J. Robinson, Ed., Robinson & Assoc., Dorset, 1975, pp. E1–E35.
42. THOMAS, G. R. and GALLAGHER, R. H., in: *Finite Elements for Thin Shells and Curved Members*, D. G. Ashwell and R. H. Gallagher, Eds., Wiley, London, 1976, pp. 155–69.
43. WASHIZU, K., *Variational Methods in Elasticity and Plasticity*, 2nd Edn, Pergamon Press, New York, 1975.
44. REISSNER, E., *J. Math. Phys.*, 29, 90–95, 1950.

45. PIAN, T. H. H., *J. AIAA*, **2**, 1333–5, 1964.
46. PIAN, T. H. H. and TONG, P., *Int. J. Num. Meth. in Engng.*, **1**, 3–28, 1969.
47. DUNGAR, R. and SEVERN, R. T., *J. Strain Analysis*, **4**, 10–21, 1969.
48. ALLMAN, D. J., *IUTAM Symp. on High Speed Computing of Elastic Structures*, University of Liège, 1971.
49. ZIENKIEWICZ, O. C., TOO, J. and TAYLOR, R. L., *Int. J. Num. Meth. in Engng.*, **3**, 275–90, 1971.
50. PAWSEY, S. F. and CLOUGH, R. W., *Int. J. Num. Meth. in Engng.*, **3**, 575–86, 1971.
51. PUGH, E. D., HINTON, E. and ZIENKIEWICZ, O. C., *Int. J. Num. Meth. in Engng.*, **12**, 1059–78, 1978.
52. HINTON, E., SCOTT, F. C. and RICKETTS, R. E., *Int. J. Num. Meth. in Engng.*, **9**, 235–56, 1975.
53. CRISFIELD, M. A., Some approximations in the non-linear analysis of rectangular plates using finite elements, *Department of the Environment, TRRL Report SR 51UC*, 1974.
54. BATOZ, J. L., BATHE, K. J. and HO, L. W., *Int. J. Num. Meth. in Engng.*, **15**, 1771–812, 1980.
55. IRONS, B. and AHMAD, S., *Techniques of Finite Elements*, Ellis Horwood Ltd, Chichester, 1980.
56. STRANG, G. and FIX, G. J., *An Analysis of the Finite Element Method*, Prentice-Hall, New Jersey, 1973.
57. FRAEIJS DE VEUBEKE, B., in: *Stress Analysis*, O. C. Zienkiewicz and G. S. Holister, Eds., Wiley, New York, 1965, Chapter 9.
58. HUGHES, T. J. R., TAYLOR, R. L. and KANOK-NUKULCHAI, W., *Int. J. Num. Meth. in Engng.*, **11**, 1529–43, 1977.
59. AHMAD, S., IRONS, B. M. and ZIENKIEWICZ, O. C., *Int. J. Num. Meth. in Engng.*, **2**, 419–51, 1973.
60. HUGHES, T. J. R. and COHEN, M., *Comp. and Struct.*, **9**, 445–50, 1978.
61. HUGHES, T. J. R., *Comp. and Struct.*, **12**, 775, 1980.
62. STRICKLIN, J. A., HAISLER, W., TISDALE, P. and GUNDERSON, R., *J. AIAA*, **7**, 180–81, 1969.
63. DHATT, G., *Proc. ASCE Symp. on Application of FEM in Civil Eng.*, Vanderbilt University, Nashville, Tennessee, 1969, pp. 13–14.
64. IRONS, B. M., in: *Finite Elements for Thin Shells and Curved Members*, D. G. Ashwell and R. H. Gallagher, Eds., Wiley, London, 1976, pp. 197–222.
65. ASHWELL, D. G. and GALLAGHER, R. H. (Eds.), *Finite Elements for Thin Shells and Curved Members*, Wiley, London, 1976.
66. GALLAGHER, R. H., *Finite Elements in Nonlinear Mechanics*, Vol. 1, Tapir, Trondheim, 1978, 243–64.
67. MORRIS, A. J., in: *Finite Elements for Thin Shells and Curved Members*, D. G. Ashwell and R. H. Gallagher, Eds., Wiley, London, 1976, pp. 15–39.
68. KOITER, W. T., in: *Theory of Inelastic Shells: First IUTAM Symp.*, Vol. 12, W. T. Koiter, Ed., North Holland, Amsterdam, 1960.
69. BUDIANSKY, B. and SANDERS, J. L. JR., in: *Progress in Applied Mechanics, The Prager Anniversary Volume*, Macmillan, New York, 1963, p. 127.
70. KOITER, W. T., *Proc. K. Ned. Akad. wet., ser. B*, **69**, 1–54, 1966.
71. SANDERS, J. L. JR., *Quarterly of Appl. Math.*, **21**, 21–36, 1963.

72. DINIS, L. M. S., MARTINS, R. A. F. and OWEN, D. R. J., in: *Numerical Methods for Nonlinear Problems*, C. Taylor *et al.*, Eds., Pineridge, Swansea, 1980, pp. 425–44.
73. BATHE, K. J. and BOLOURCHI, S., *Comp. and Struct.*, **11**, 23–48, 1980.
74. KRAKELAND, B., in: *Finite Elements in Nonlinear Mechanics*, Vol. 1, P. Bergan *et al.*, Eds., Tapir, Trondheim, 1978, pp. 265–84.
75. JAVAHERIAN, H., DOWLING, P. J. and LYONS, L. P. R., *Comp. and Struct.*, **12**, 147–59, 1980.
76. KANOK-NUKULCHAI, W., TAYLOR, R. L. and HUGHES, T. J. R., in: *Computational Methods in Nonlinear Structural and Solid Mechanics*, A. K. Noor and H. G. McComb, Jr., Eds., Pergamon Press, Oxford, 1981, pp. 19–27.
77. RAMM, E., in: *Formulations and Computational Algorithms in Finite Element Analysis*, K. J. Bathe, J. T. Oden and W. Wunderlich, Eds., MIT Press, Cambridge MA, 1977, Chapter 10.
78. BRENDEL, B. and RAMM, E., in: *Int. Conf. on Engng. Appl. of the Finite Element Method*, Vol. 2, Computas, Hovik, Norway, 1979, 18.1–18.30.
79. ZIENKIEWICZ, O. C., PAREKH, C. J. and KING, I. P., *Proc. Symp. Arch Dams*, *Inst. Civ. Eng.*, London, 1968.
80. HORRIGMOE, G. and BERGAN, P. G., *Comp. Meth. Appl. Mech. and Engng*, **16**, 11–35, 1978.
81. CLOUGH, R. W. and JOHNSON, C. P., *Int. J. Solids and Struct.*, **4**, 43–60, 1968.
82. CLOUGH, R. W. and TOCHER, J. L., *Proc. Symp. on Theory of Arch Dams*, *Southampton University, 1964*, Pergamon Press, Oxford, 1965.
83. ARGYRIS, J. H., DUNNE, P. C., MALEJANNAKIS, G. A. and SCHELKE, E., *Comp. Meth. Appl. Mech. and Engng.*, **10**, 341–403., 1977.
84. CONNOR, J. and BREBBIA, C., *Proc. ASCE, Eng. Mech. Div.*, **93**, 43–65, 1967.
85. COWPER, G. R., LINDBERG, G. M. and OLSON, M. D., *Int. J. Solids and Struct.*, **6**, 1135–56, 1970.
86. STRICKLAND, C. E. and LODEN, W. A., *Proc. 2nd Comp. Matrix Methods in Struct. Mech.*, *AFFDL-TR-68-150*, Wright Patterson Air Force Base, Ohio, 1968, pp. 641–66.
87. MORRIS, A. J., *Int. J. Solids and Struct.*, **9**, 331–45, 1973.
88. VLASOV, V. Z., General theory of shells and its application in engineering, *NASA TTF-99*, 1964.
89. BELYTSCHKO, T. and GLAUM, L. W., *Comp. and Struct.*, **10**, 175–82, 1979.
90. CRISFIELD, M. A., Linear and nonlinear finite element analysis of cylindrical shells, *Department of the Environment, TRRL Report LR987*, 1981.
91. BRODLIE, K. W., in: *Unconstrained Minimisation, the State of the Art in Numerical Analysis*, *I.M.A.*, D. A. D. Jacobs, Ed., Academic Press, New York, 1977.
92. WOLFE, M. A., *Numerical Methods for Unconstrained Optimisation—An Introduction*, Van Nostrand Rheinhold, London, 1978.
93. CRISFIELD, M. A., Iterative solution procedures for linear and nonlinear structural analysis, *Department of the Environment, TRRL Report LR 900*, 1979.
94. JENNINGS, A., *Matrix Computation for Engineers and Scientists*, Wiley, New York, 1977.

95. JENNINGS, A., *Comp. J.*, **9**, 281–5, 1966.
96. IRONS, B. M., *Int. J. Num. Meth. in Engng.*, **2**, 5–32, 1972.
97. HESTENES, M. and STEIFEL, E., *J. Res. of Nat. Bur. of Standards*, **49**(6), 403–36, 1952.
98. FLETCHER, R. and REEVES, C. M., *Comp. J.*, **7**, 149–254, 1964.
99. POLAK, E. and RIBIÈRE, G., Note sur la convergence de méthodes de directions conjugées, *Revue Française Inform. Rech. Operations*, *16-R1*, 35–43, 1969.
100. POWELL, M. J. D., *Math. Program*, **12**, 241–54, 1977.
101. SHANNO, D. F., *Math. of O.R.*, **13**(3), 244–55, 1978.
102. JENNINGS, A. and MALIK, G. M., *Int. J. Num. Meth. in Engng.*, **12**, 141–58, 1978.
103. MEIERINK, J. A. and VON DER VORST, H. A., *Math. of Comp.*, **31**, 148–62, 1977.
104. KERSHAW, D. S., *J. Comp. Phys.*, **26**, 43–65, 1978.
105. FOX, R. L. and STANTON, E. L., *J. AIAA*, **6**(6), 1036–42, 1968.
106. OTTER, J. R. H. and DAY, A. S., *The Engineer*, **209**, 177–82, 1960.
107. CASSELL, A. C., *Proc. Instn. Civ. Engrs.*, **45**, 65–78, 1970.
108. FRANKEL, S. P., *Maths Tables Aids Comput.*, **4**, 65–75, 1950.
109. LYNCH, R. D., KELSEY, S. and SAXE, H. C., The application of D. R. to the finite element method of structural analysis, *Tech. Report No. THEMIS-UND-68-1*, University of Notre Dame, September 1968.
110. CRISFIELD, M. A., *Meth. in App. Mech. and Engng.*, **20**, 1979, 267–78, 1979.
111. IRONS, B. and ELSAWAF, A., in: *Formulations and Algorithms in Finite Element Analysis*, K. J. Bathe, J. T. Oden and W. Wunderlich, Eds., MIT Press, Cambridge MA, 1977, pp. 656–72.
112. CRISFIELD, M. A., in: *Numerical Methods for Non-Linear Problems*, Vol. 1, C. Taylor *et al.*, Eds., Pineridge, Swansea, 1980, pp. 261–90.
113. KEY, S. W., STONE, C. M. and KRIEG, R. D., *Europe/US Workshop on Nonlinear Finite Element Analysis in Structural Mechanics*, Bochum, West Germany, July 1980.
114. UNDERWOOD, P., in: *Computational Methods for Transient Response*, T. Belytschko and T. J. R. Hughes, Eds., North Holland, Amsterdam, in press.
115. PAPADRAKAKIS, M., *Comp. Meth. Appl. Mech. and Engng.*, **25**, 35–48, 1981.
116. DAVIDON, W. C., Variable metric method for minimisation, *Argonne Nat. Lab. Report ANL-5990*, 1959.
117. FLETCHER, R. and POWELL, M. J. D., *Comp. J.*, **6**, 163–8, 1963.
118. DENNIS, J. JR. and MORE, J., *SIAM Review*, **19**(1), 46–84, January 1977.
119. BROYDEN, C. G., *J. Inst. Math. Appl.*, **6**, 222–31, 1970.
120. FLETCHER, R., *Comp. J.*, **13**, 317–22, 1970.
121. LITTLE, G. H., *Int. J. Mech. Sci.*, **19**, 725–43, 1977.
122. DAWE, D. J., *Int. J. Num. Meth. in Engng.*, **3**, 589–93, 1971.
123. MATTHIES, H. and STRANG, G., *Int. J. Num. Meth. in Engng.*, **14**, 1613–26, 1979.
124. BATHE, K. J. and CIMENTO, A. P., *Comp. Math. Appl. Mech. and Engng.*, **22**, 59–86, 1980.
125. ABDEL RAHMAN, H. H., HINTON, E. and HUQ, M. M., in: *Numerical Methods for Nonlinear Problems*, Vol. 1, C. Taylor *et al.*, Eds., Pineridge, Swansea, 1980, pp. 493–9.

126. GERADIN, M., IDELSOHN, S. and HOGGE, M., in: *Computational Methods in Nonlinear Structural and Solid Mechanics*, A. K. Noor *et al.*, Eds., Pergamon Press, New York, 1980, pp. 73–82.
127. TOINT, P. L., *Math. Comp.*, **31**, 954–61, 1977.
128. NOOR, A. K., *Comp. and Struct.*, **13**, 31–44, 1981.
129. NAGY, D. A. and KÖNIG, M., *Comp. Meth. Appl. Mech. and Engng.*, **19**, 447–84, 1979.
130. KOITER, W. T., On the stability of elastic equilibrium, *Thesis*, Delft, H. J. Paris, Amsterdam, 1945 (in Dutch). English translation: Wright Patterson Air Force Base, Ohio, Rep. No. AFF DL-TR-70-25, 1970.
131. KOITER, W. T., *Proc. Symp. Nonlinear Probl.*, University of Wisconsin Press, Madison, 1963, pp. 257–75.
132. BESSELING, J. F., *Comp. Meth. Appl. Mech. and Engng.*, **3**, 173–94, 1974.
133. CARNOY, E., *Comp. Meth. Appl. Mech. and Engng.*, **23**, 143–74, 1980.
134. HAFTKA, R. T., MALLET, R. H. and NACHBAR, W., *Int. J. Solids and Struct.*, **7**, 1427–45, 1971.
135. THOMPSON, J. M. T. and WALKER, A. C., *Int. J. Solids and Struct.*, **4**, 757–68, 1968.
136. WALKER, A. C., *Int. J. Num. Meth. in Engng.*, **1**, 177–80, 1969.
137. HACKBUSCH, W., *Num. Meth. for Nonlinear Problems*, C. Taylor *et al.*, Eds., Pineridge, Swansea, 1980, pp. 1041–50.
138. WACHSPRESS, E. L., in: *Formulations and Algorithms in Finite Element Analysis*, K. J. Bathe, J. T. Oden and W. Wunderlich, Eds., MIT Press, Cambridge MA, 1977, pp. 877–913.
139. PEANO, A., *Comp. and Math. with Appls.*, **2**, 211–24, 1976.
140. BERGAN, P. G. and SOREIDE, T., *Finite Elements in Nonlinear Mechanics*, Tapir, Trondheim, 1978, pp. 647–69.
141. BERGAN, P. G., HORRIGMOE, G., KRAKELAND, B. and SOREIDE, T. H., *Int. J. Num. Meth. in Engng.*, **12**, 1677–96, 1978.
142. CRISFIELD, M. A., *Proc. Instn. Civ. Engrs.*, *Part 2*, **69**, 891–909, 1980.
143. RAMM, E., *Europe/US Workshop on Nonlinear Element Analysis in Structural Mechanics*, Bochum, West Germany, July 1980.
144. CRISFIELD, M. A., in: *Computational Methods in Nonlinear Structural and Solid Mechanics*, A. K. Noor and H. G. McComb Jr., Eds., Pergamon Press, Oxford, 1980, pp. 55–62. (Also in: *Comp. and Struct.*, **13**, 55–62, 1981.)
145. RIKS, E., *Int. J. Solids and Struct.*, **15**, 529–51, 1979.
146. RIKS, E., *J. Appl. Mech.*, **39**, 1060–66, 1972.
147. WEMPNER, G. A., *Int. J. Solids and Struct.*, **7**, 1581–99, 1971.
148. RIKS, E., *Acta Technica Scientiarum Hungaricae*, *Tomus*, **87**, 121–41, 1978.
149. OWEN, D. R. J. and HINTON, E., *Finite Elements in Plasticity: Theory and Practise*, Pineridge, Swansea, 1980.
150. YAMADA, Y., YOSHIMUKA, N. and SAKURAI, T., *Int. J. Mech. Sci.*, **10**, 343–54, 1968.
151. ZIENKIEWICZ, O. C., VALLIAPAN, S. and KING, I. P., *Int. J. Num. Meth. in Engng.*, **1**, 75–100, 1969.
152. NAYAK, G. C. and ZIENKIEWICZ, O. C., *Int. J. Num. Meth. in Engng.*, **5**, 113–35, 1972.
153. BUSHNELL, D., *Int. J. Num. Meth. in Engng.*, **11**, 683–708, 1977.

154. ARGYRIS, J. H., VAZ, L. E. and WILLAM, K. J., *Comp. Meth. Appl. Mech. and Engng.*, **16**, 231–77, 1978.
155. TRACEY, D. M. and FREESE, C. E., *Comp. and Struct.*, **13**, 45–53, 1981.
156. ARMEN, H., PIFKO, A. B., LEVINE, H. S. and ISAKSON, G., *Finite Element Techniques in Structural Mechanics*, H. Tottenham and C. Brebbia, Eds., Southampton University Press, 1970, Chapter 8.
157. MARCAL, P. V. and PILGRIM, W. R., *J. Strain Analysis*, **1**(4), 339–50, 1966.
158. CRISFIELD, M. A., Large-deflection elasto-plastic buckling analysis of plates using finite elements, *Department of the Environment, TRRL Report LR 593*, 1973.
159. CORMEAU, I., *Int. J. for Num. Meth. in Engng.*, **12**, 203–28, 1978.
160. CRISFIELD, M. A., Ivanov's yield criterion for thin plates and shells using finite elements, *Department of the Environment, TRRL Report LR 919*, 1979.
161. KRÖPLIN, B. H. and DUDDECK, H., Simplified calculation models applied to post buckling analysis of thin plates. *Europe/US Workshop on Nonlinear Finite Element Analysis in Structural Mechanics*, Bochum, West Germany, July 1980.
162. WEMPNER, G. A. and HWANG, C. M., *Int. J. Solids and Struct.*, Vol. **16**, 161–5, 1980.
163. EIDSHEIM, O. M. and LARSEN, P. K., *Europe/US Workshop on Nonlinear Finite Element Analysis in Structural Mechanics*, Bochum, West Germany, July 1980.
164. CRISFIELD, M. A. and PUTHLI, R. S., *Finite Elements in Nonlinear Mechanics*, Vol. 1, Tapir, Trondheim, 1978, pp. 373–92.
165. PUTHLI, R. S., CRISFIELD, M. A. and SUPPLE, W. J., Stability of steel structures, *Preliminary Report IABSE, Liège*, 1977, pp. 427–31.
166. SCORDELIS, A. C., *The Finite Element Method in Civil Engineering*, McGill University, Montreal, 1972.
167. NILSON, A. H., *ACI J.*, **65**, 152–63, 1968.
168. BLAAUWENDRAAD, J. and GROOTENBOER, H. T., *IABSE Colloquium on Advanced Mechanics of Reinforced Concrete—Working Papers*, Delft, 1981, pp. 442a–422j.
169. PHILLIPS, D. V. and ZIENKIEWICZ, O. C., *Proc. Instn. Civ. Engrs.*, Part 2, **61**, 59–88, 1976.
170. BERGAN, P. G. and HOLAND, I., *Comp. Meth. Appl. Mech. and Engng.*, **17/8**, 443–67, 1979.
171. BAZANT, Z. P., *IABSE Colloquium on Advanced Mechanics of Reinforced Concrete—Introductory Report*, Delft, 1981, pp. 9–39.
172. ARGYRIS, J. H., FAUST, G. and WILLAM, K. J., *IABSE Colloquium on Advanced Mechanics of Reinforced Concrete—Introductory Report*, Delft, 1981, pp. 85–106.
173. COPE, R. J., RAO, P. V. and EDWARDS, K. R., in: *Numerical Methods for Nonlinear Problems*, Vol. 1, C. Taylor *et al.*, Eds., Pineridge, Swansea, 1980, pp. 445–70.
174. ABDEL RAHMAN, H. H., HINTON, E. and HUQ, M. M., in: *Numerical Methods for Nonlinear Problems*, Vol. 1, C. Taylor *et al.*, Pineridge, Swansea, 1980, pp. 493–9.

175. LIN, C. S. and SCORDELES, A. C., *J. Struct. Div. Am. Soc. Civ. Engrs.*, **101**(ST3), 523–38, 1975.
176. HAND, F. A., PECKNOLD, D. A. and SCHNOBRICH, W. C., *J. Struct. Div. Am. Soc. Civ. Engrs*, **99**(ST7), 1491–505, 1973.
177. GILBERT, R. I. and WARNER, R. F., *J. Struct. Div. Am. Soc. Civ. Engrs.*, **104**(ST12), 1885–900, 1976.
178. CRISFIELD, M. A., *Fenomech 81*, Stuttgart, August 1981.
179. KUPFER, H., HILSDORF, H. K. and RUSCH, H., *J. Am. Conc. Inst.*, **66**, 656–66, 1969.
180. FRANCHI, A. and COHN, M. Z., *Comp. Struct.*, **11**, 421–7, 1980.
181. DUDDECK, H., GRIEBENOU, G. and SCHAPER, G., *Nonlinear Behaviour of Reinforced Concrete Spatial Structures—Preliminary Report 1*, G. Mehlhorn *et al.*, Eds., Werner Verlag, Dusseldorf, 1978, pp. 101–13.
182. HUGHES, T. J. R., *Appl. Mech. Rev.*, **33**(11), 1467–77, 1980.

INDEX